Se94

The Rand McNally Library of Astronomical Atlases
for Amateur and Professional Observers
Series Editors Garry Hunt and Patrick Moore

# The Sun

## Iain Nicolson

Foreword by Professor Archie E. Roy

Published in Association with
the Royal Astronomical Society

Rand McNally and Company
New York · Chicago · San Francisco

**The Sun**
© Mitchell Beazley Publishers 1982
© Text: Iain Nicolson 1982

ISBN 528-81540-7
Library of Congress Catalog Number 80-53882

**The Sun** was edited and designed by
Mitchell Beazley Publishers,
Mill House, 87–89 Shaftesbury Avenue,
London W1V 7AD

Phototypeset by Servis Filmsetting Ltd.
Origination by Adroit Photo Litho Ltd.
Printed in the United States

**Editor** Gilead Cooper
**Designer** Sean Keogh
**Editorial Assistant** Charlotte Kennedy
**Picture Research** Meg Price-Whitlock

**Executive Editor** Lawrence Clarke
**Executive Art Editor** John Ridgeway
**Production Manager** Barry Baker

The units and notation used throughout this book are based on the Système International des unités (SI units), which is currently being introduced universally for scientific and educational purposes. There are seven "base" units in the system: the *meter* (m), the *kilogram* (kg), the *second* (s), the *ampere* (A), the *kelvin* (K), the *mole* (mol) and the *candelo* (cd). Other quantities are expressed in units derived from the base units; thus, for example, the unit of force the newton (N) is defined as the force required to give a mass of one kilogram an acceleration of one meter per second squared ($kg\,m\,s^{-2}$).

Some branches of science continue to adhere to a few of the older units, and in one case an editorial concession has had to be made to existing scientific usage: the SI unit of magnetism, the tesla, has been dropped in favor of the more common unit, the gauss. One tesla is equal to 10,000 gauss.

For very large and very small numbers, "index notation" has been adopted, so that where appropriate numbers are written as powers of ten. For example, 1,000,000 may be written as $10^6$, and 3,500,000 as $3.5 \times 10^6$. Numbers smaller than one are indicated by negative powers: thus 0.00035 is written as $3.5 \times 10^{-4}$. In addition a variety of prefixes is used to denote certain multiples of units (*see* Table 1). Table 2 gives the SI equivalents of common imperial units while Table 3 lists a selection of astronomical constants.

**Table 1: SI prefixes**

| Factor | Name | Prefix Symbol |
|---|---|---|
| $10^{18}$ | exa | E |
| $10^{15}$ | peta | P |
| $10^{12}$ | tera | T |
| $10^{9}$ | giga | G |
| $10^{6}$ | mega | M |
| $10^{3}$ | kilo | k |
| $10^{2}$ | hecto | h |
| $10^{1}$ | deca | da |
| $10^{-1}$ | deci | d |
| $10^{-2}$ | centi | c |
| $10^{-3}$ | milli | m |
| $10^{-6}$ | micro | $\mu$ |
| $10^{-9}$ | nano | n |
| $10^{-12}$ | pico | p |
| $10^{-15}$ | femto | f |
| $10^{-18}$ | atto | a |

**Table 2: SI conversion factors**

| | |
|---|---|
| **Length** | |
| 1 in | 25.4 mm |
| 1 mile | 1.609344 km |
| **Volume** | |
| 1 imperial gal | 4.54609 cm³ |
| 1 US gal | 3.78533 liters |
| **Velocity** | |
| 1 ft/s | $0.3048\,m\,s^{-1}$ |
| 1 mile/h | $0.44704\,m\,s^{-1}$ |
| **Mass** | |
| 1 lb | 0.45359237 kg |
| **Force** | |
| 1 pdl | 0.138255 N |
| **Energy (work, heat)** | |
| 1 cal | 4.1868 J |
| **Power** | |
| 1 hp | 745.700 W |
| **Temperature** | |
| °C | = kelvins − 273.15 |
| °F | = $\frac{9}{5}$ ( C) + 32 |

**Table 3: Astronomical and physical constants**

| | |
|---|---|
| **Astronomical unit (A.U.)** | $1.49597870 \times 10^8$ km |
| **Light-year (l.y.)** | $9.4607 \times 10^{12}$ km = 63,240 A.U. = 0.306660 pc |
| **Parsec (p.c.)** | $30.857 \times 10^{12}$ km = 206,265 A.U. = 3.2616 l.y. |
| **Length of the year** | |
| Tropical (equinox to equinox) | $365^d.24219$ |
| Sidereal (fixed star to fixed star) | 365.25636 |
| Anomalistic (apse to apse) | 365.25964 |
| Eclipse (Moon's node to Moon's node) | 346.62003 |
| **Length of the month** | |
| Tropical (equinox to equinox) | $27^d.32158$ |
| Sidereal (fixed star to fixed star) | 27.32166 |
| Anomalistic (apse to apse) | 27.55455 |
| Draconic (node to node) | 27.21222 |
| Synodic (New Moon to New Moon) | 29.53059 |
| **Length of day** | |
| Mean solar day | $24^h03^m56^s.555 = 1^d.00273791$ mean solar time |
| Mean sidereal day | $23^h56^m04^s.091 = 0^d.99726957$ mean solar time |
| Earth's sidereal rotation | $23^h56^m04^s.099 = 0^d.99726966$ mean solar time |
| **Speed of light in vacuo (c)** | $2.99792458 \times 10^5\,km\,s^{-1}$ |
| **Constant of gravitation** | $6.672 \times 10^{-11}\,kg^{-1}\,m^3\,s^{-2}$ |
| **Charge on the electron (e)** | = 1.602 coulomb |
| **Planck's constant (h)** | $= 6.624 \times 10^{-34}$ J s |
| **Solar radiation** | |
| Solar constant | $1.39 \times 10^3\,J\,m^{-2}\,s^{-1}$ |
| Radiation emitted | $390 \times 10^{26}\,J\,s^{-1}$ |
| Visual absolute magnitude ($M_v$) | + 4.79 |
| Effective temperature | 5,800 K |

# Contents

# Foreword

Astronomy is the oldest of the sciences, born of the fact that man evolved on a planet from which he could see the sky. For 50,000 years he has had intelligence enough to study the heavens and attempt to understand what he saw by day and by night. And there is no doubt that what he saw and deduced from his observations has had extraordinary effects upon his life in many lands.

If our civilization had developed in the way it has on a planet whose skies were eternally cloud-covered (which is doubtful), we would have believed, up to some forty years ago, that the Earth *was* the universe. Only with the advent of large radar dishes, high-flying aircraft and rockets would the shattering fact have emerged that above the opaque cloud-layer lay a seemingly boundless universe.

Modern western civilization has been greatly influenced by Copernicus, Kepler, Galileo and Newton, all watchers of the skies. Our belief in a rational universe capable of being understood and our scientific and technological civilization spring from the cyclic behaviour of Sun, Moon, planets and stars and Newton's ability to explain so much of that behaviour by his law of gravitation and his three laws of motion. Timekeeping, navigation, geodesy, dynamics, religious and philosophical systems, cosmology and relativity and many other activities and interests of man have been directly affected by our study of the heavens.

There have been three astronomical revolutions. The first—the serious, naked-eye study of the heavens—lasted a long time—at least five millennia—and ended when Galileo began his systematic telescopic study of the sky in AD 1610. That second revolution, in which the camera and the spectroscope played their part, was brought to a climax by the enormous amount of information gathered by telescopes such as the 200-inch Hale telescope at Mt Palomar. On October 4, 1957, with the orbiting of Sputnik I, the third astronomical revolution began. Not only can we now place instruments in orbit above the Earth's atmosphere, obtaining access to the entire electromagnetic spectrum, but we can send spacecraft such as the Mariners, Pioneers and Voyagers to other planets in our Solar System. The flood of astronomical information has become a torrent, sweeping away many of our former ideas about the universe.

The time therefore seemed ripe for a series of atlases designed to take stock of this flood of new information and the new understanding it has brought us of the nature of the universe. Each atlas in the series has been written by an author chosen by Mitchell Beazley Publishers so that the text will provide the most up-to-date assessment of the celestial body studied, together with explanatory diagrams and the most modern pictures. Each author's text has then been carefully checked and authenticated by an acknowledged expert in the subject chosen by the Royal Astronomical Society's Education Committee. The final text of each book should therefore truly convey our present-day knowledge of the subject and remain a definitive work for many years to come.

**Archie E. Roy**
BSc, PhD, FRAS, FRSE, FBIS
Titular Professor of Astronomy in the
University of Glasgow
Chairman of the Education Committee of the
Royal Astronomical Society

# Introduction

The Sun is of vital importance to us. Without the Sun there would be no life on Earth. Our nearest star, it is the primary source of light, heat and energy for our planet, and its crucial significance was recognized by all early civilizations. The Sun was regarded as a deity. To the ancient Egyptians, for example, the Sun god, Ra, was born each morning and carried across the heavens in a boat to die in the west each evening; during the night he was carried beneath the world to his rebirth in the east. Or again, to the ancient Greeks he was Apollo, riding across the heavens in a fiery chariot.

Yet even in early Greek times, some 2,500 years ago, there were those who speculated about the physical nature of this brilliant orb, and attempts were made to gauge its size and distance. By the third century BC there were some who were prepared to suggest that the planets, and perhaps even the Earth, moved around the Sun. However, it was not until the seventeenth century that the idea of a Sun-centered universe gained widespread acceptance. In 1610 telescopic observations of the Sun were made and the modern era of solar astronomy began.

With the passage of time we have come to realize that the Sun is an ordinary star occupying no particularly significant place in the Galaxy; it shines in the same way as the other stars are believed to shine, and it is only because of its close proximity to the Earth that the Sun is exceptionally important to us.

From the astronomer's point of view the Sun is of the utmost interest because it is the only star that can be studied in detail. Until comparatively recently the Sun could be studied only by means of the visible light emitted from its surface. Then came the advent of radio astronomy, which revealed many new facets of solar structure and activity. Although modern techniques have permitted other radiations to be detected to a limited degree from Earth-based observatories and high-altitude balloons, it was the development of space technology that suddenly opened up new vistas for the solar physicist; as a result the past two decades have revolutionized our understanding of the Sun.

Satellites in orbit around the Earth can study the Sun 24 hours a day over the widest range of wavelengths and can detect radiations that cannot penetrate our atmosphere to reach ground level. Most spectacular in this respect has been the new detail and structure revealed at X-ray and ultraviolet wavelengths. Spacecraft exploring interplanetary space have made a study of solar particle emissions and the interplanetary magnetic field which simply could not have been achieved by ground-based astronomers. The manned missions to the Skylab laboratory during 1973 and 1974 contributed a wealth of new data and spectacular photography which served to bring the new view of the Sun into the public eye in the most dramatic fashion; but there has been a multitude of unmanned spacecraft—most recently the Solar Maximum Mission—which has served to shape our current understanding, and we look forward with anticipation to further exciting discoveries from new generations of solar satellites in the Space Shuttle era.

Ground-based instruments continue to make a major contribution; indeed detectors buried down mineshafts and within mountains are searching for elusive particles, neutrinos, emitted from the core of the Sun. The results of experiments such as these are throwing up challenges to concepts that we had hitherto regarded as well established. The analysis of historical records, too, is throwing up many surprises. Links between changing levels of activity on the Sun and climatic changes here on Earth are beginning to emerge, and there is mounting evidence to suggest that the Sun may not be quite so constant a star as we thought. It is an exciting time for solar physics.

Yet despite the great concentration of professional effort devoted to the study of the Sun, it remains a fruitful object of study for the amateur observer, and it is still possible for the amateur to make a contribution to advancing our understanding of this, the most vital of stars.

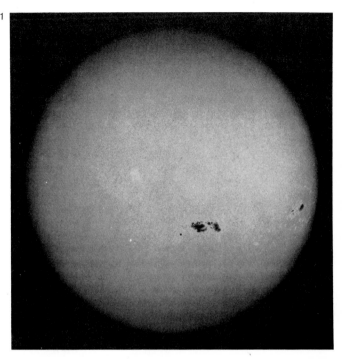

**1. The Sun**
White light photograph taken on 4 July 1974. Kitt Peak National Observatory, Arizona.

**2. Symbol of the Sun**

**3. Apollo Belvedere, Vatican Museum**

5

# The Sun as a Star

The Sun is an ordinary, middle-aged star. It is believed to be at least $4.6 \times 10^9$ years old, and to be about half-way through its stable life. There are many stars that are older, and many that are younger. In fact there appears to be nothing unique or distinctive about the Sun, and it only seems especially significant from Earth because it is so much nearer than any other star. As a source of light, heat and energy the Sun is of vital importance to the Earth; without it, human life would not exist.

The Sun is a member of a huge system or "Galaxy" of stars consisting of a central hub of stars surrounded by a relatively thin disc made up of stars together with clouds of gas and dust. The entire system contains about $10^{11}$ stars, and measures approximately 30,000 parsecs in diameter (the parsec being a unit of distance equal to 3.26 light-years—*see* Glossary). The mean separation of stars in the galactic disc is about 1 parsec; the Sun's nearest neighbor, a dull red star called Proxima Centauri, lies at a distance of 1.3 parsecs. The Sun itself is located in the galactic suburbs, some 9,000 parsecs from the center, which it revolves around at a mean speed of about $250 \, \text{km s}^{-1}$, taking some 225 million years to complete each circuit; this period of time is known as a "cosmic year".

As well as sharing in the general rotation of the Galaxy, the Sun, together with its planets and other attendant bodies, is moving relative to the mean motion of the neighboring stars at a speed of $19.7 \, \text{km s}^{-1}$. Observations over a long period of time have revealed small changes in the positions of the stars. These changes are largely due to the individual motions of the stars themselves, but there is in addition a general tendency to diverge from a point in the constellation Hercules, which indicates that the Sun is in fact heading towards that point. The direction in which the Sun is moving is known as the "solar apex"; the opposite direction is the "solar antapex".

Like other stars, the Sun is a self-luminous globe of incandescent gas, generating energy by means of nuclear reactions taking place deep within its interior (*see* pages 82–83). Its principal constituents are hydrogen (making up about 73 percent of its mass) and helium (about 25 percent), the remaining chemical elements making up rather less than 2 percent of the total. Whereas the Earth contains a high proportion of metals, silicates and other relatively heavy minerals, the Sun has a composition that reflects the general distribution of the chemical elements in the universe.

The Sun is the only star whose surface can be seen and studied directly. The other stars are so far distant that even with the largest Earth-based telescopes, they appear only as points of light; telescopes render the images of stars much brighter, but they do not show the stars as visible discs. Small telescopes will show details on the Sun's surface, but it cannot be emphasized too strongly that it is dangerous to stare at the Sun even without optical aid, and *on no account should one look directly at the Sun* through binoculars or a telescope. Serious and irreparable eye damage, if not total blindness, would be the result of such an act. The simplest way of observing the Sun safely with the aid of a telescope is by projecting the image onto a sheet of smooth white paper or card held at a distance behind the eyepiece.

## Magnitude and energy

The brilliance of the Sun overwhelms that of the other stars in the daylight sky. The apparent brightness of stars is measured on the stellar magnitude scale, a cumbersome system whereby the *lower* the numerical value of magnitude, the brighter the star. A bright star, such as Antares (in the constellation of Scorpius) or Spica (in Virgo), has an apparent magnitude of about $+1$, while the faintest star visible to the unaided eye under ideal conditions is about magnitude $+6$. Stars too faint to be seen without telescopic aid have higher values of magnitude, while the brightest stars of all have negative values; for example, the brightest star in the sky,

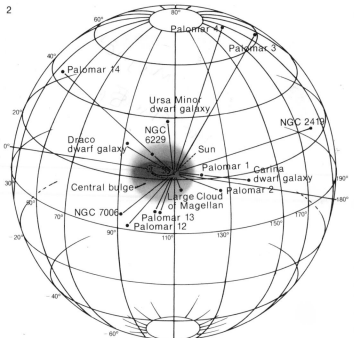

**1. Panorama of the Milky Way**
The band of faint light visible in the sky on a clear night is known as the "Milky Way". The effect is the result of viewing the Galaxy, which is a flattened disc in shape, edge on from inside.

**2. The Galaxy**
The principal features of the Galaxy are a dense central bulge and fairly loose spiral arms; these are surrounded by a halo of material. There is very great uncertainty about the overall size and mass of the Galaxy: recent evidence suggests it may be surrounded by a "corona" of material (including dwarf galaxies and globular clusters) which extends out to a distance of some 100,000 pc.

## 3. Magnitude scale

The scale of stellar magnitudes is based on a system originally devised by Hipparchus in the second century BC. The brightest stars were said to be of the first magnitude, while the faintest stars visible to the unaided eye were said to be of the sixth magnitude. It was later discovered that this range of brightnesses represented a factor of about 100: thus the difference between a given magnitude and the next in scale represents an increase in brightness by a factor of $\sqrt[5]{100}$. The lower the value of the magnitude, the brighter the star. This system has been extended to include both very faint objects and very bright ones (which are assigned negative magnitudes).

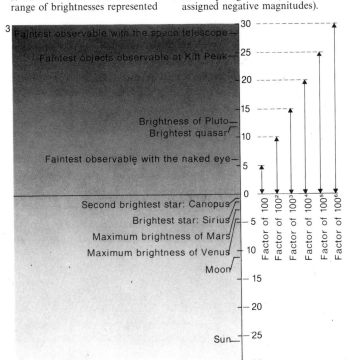

Sirius, has an apparent magnitude of $-1.4$. The apparent magnitude of the Sun is $-26.7$. Since each step of five magnitudes corresponds to a factor of 100 in brightness, it follows that from Earth the Sun appears brighter than Sirius by a factor in excess of 10,000 million. The magnitude difference reflects the response of the human eye, compressing a large dynamic range.

The apparent magnitude of a star is a measure of the amount of light arriving from it and depends on a number of factors, notably the star's "luminosity" (its output of light and other radiation) and its distance from the Earth. The apparent brightness of a source diminishes with the square of its distance. The absolute magnitude of a star, on the other hand, is defined as the apparent magnitude the star would have if it were located at a standard distance of 10 parsecs: absolute magnitude thus provides a measure of the relative luminosities of stars located at different distances. The absolute magnitude of the Sun is $+4.8$, so that if it were located at a distance of 10 parsecs from Earth—which is relatively close on the astronomical scale—it would be a rather insignificant star which probably would not be visible in urban night skies.

The luminosity of a star is the total amount of energy it emits per second, its "power output", and this energy is emitted in the form of radiation of all kinds; in common with many other stars, the Sun emits a large fraction of its energy in the form of visible light. The solar luminosity is $3.86 \times 10^{26}$ watts ($J s^{-1}$), a prodigious outpouring of energy of which the entire Earth intercepts only about four parts in 10,000 million. There are many stars which are far more luminous than the Sun, but a greater proportion are less luminous. Denoting the luminosity of the Sun by $L_\odot$, it is found that most stars have values between about $10^4 L_\odot$ and $10^{-4} L_\odot$, although there are stars with luminosities as great as $10^6 L_\odot$ or as little as $10^{-5} L_\odot$. The Sun lies at about the geometric mean of the range.

Some stars, known as variable stars, fluctuate in brightness by large amounts—some in a regular, periodic way, others in an irregular fashion. The output of solar radiation seems to have been remarkably constant, on average, over the past few thousand million years, but recently it has become apparent that the Sun is active in a variety of ways, and that changes in its activity have significant effects here on Earth. The growth of interest in solar research reflects in part an increasing awareness of this link.

The effective surface temperature of most stars lies between about 30,000 K and about 3,000 K—that of the Sun being 5,780 K—although there are exceptions at each end of the range. There are more stars cooler than the Sun than there are hotter.

### Mass and density

The Sun has a mass of $1.9891 \times 10^{30}$ kg (denoted by $M_\odot$). This value lies near the middle of the stellar range, most stars having masses of between $10 M_\odot$ and $0.1 M_\odot$. The most massive stars of all may be as great as $100 M_\odot$, while the lower limit for a star capable of shining by means of internal nuclear reactions is considered to be about $0.085 M_\odot$. Stellar diameters and densities span huge ranges. Stars known as red giants and supergiants (*see* pages 86–87) may be hundreds of times larger than the Sun, while compact stars, known as white dwarfs, are comparable in size with the Earth, having about one hundredth of the solar diameter. The relatively rarely detected neutron stars have radii of only about 10 km.

Mean densities for the largest stars are appreciably less than that of air at ground level, while white dwarfs are so dense that a centimeter cube of their material would if brought to Earth weigh about 1 tonne. Neutron stars are about 1,000 million times denser again. The density of the Sun falls between these extremes, being equal to $1.41 \ 4 10^3$ kg m$^{-3}$ or just under one and a half times the density of water. However, it should be noted that the very low mean densities of some stars are due to the extended nature of their tenuous outer atmospheres, whereas their central densities are comparable with the Sun.

# The Solar System

The Solar System comprises the Sun, the nine major planets (Mercury, Venus, Earth, Mars, Jupiter, Saturn, Uranus, Neptune and Pluto) together with their satellites, a host of minor bodies (the asteroids, comets and meteoroids) and a certain amount of interplanetary gas and dust. The Sun is by far the dominant body, its gravitational field controlling the motions of the other bodies in the system; the solar magnetic field and solar radiation play an important role, too, particularly with regard to the gas and dust.

The Sun lies at a mean distance from the Earth of 149,600,000 km, a distance known as the astronomical unit (A.U.). With a diameter of 1,392,000 km, the Sun is 109 times larger than the Earth and nearly 10 times the diameter of Jupiter, the largest of the planets. Its mass—$1.9891 \times 10^{30}$ kg—is some 330,000 times that of the Earth, and over 1,000 times that of Jupiter; indeed, the Sun contains over 99.8 percent of the mass of the Solar System. When its volume, 1,303,000 times that of the Earth, is compared with its mass, it is clear that the mean density of the Sun is appreciably less than the Earth's density. In fact, the mean density of solar material ($1,410 \text{ kg m}^{-3}$) is just under one quarter of the value for terrestrial material, and this reflects the difference in structure and composition of the two bodies.

## Kepler's laws

It was discovered by Johannes Kepler in 1609 that the planets travelled around the Sun in ellipses, with the Sun located at one focus of the ellipse. This principle is the first of the three "laws" governing planetary motion that bear Kepler's name. Strictly speaking, the Sun and planets should be considered to revolve around the center of mass or "barycenter" of the entire Solar System, a point which does not coincide exactly with the center of the Sun. However, since the total mass of the nine planets amounts to less than 0.2 percent of the mass of the Sun, the discrepancy may, for most purposes, be ignored.

Kepler's second law states that for each planet the line between the planet and the Sun (the "radius vector") sweeps out equal areas in equal times. When a planet's orbit takes it close to the Sun it must therefore move faster than when it is far away. The point of closest approach to the Sun is called the "perihelion", and the most distant point is the "aphelion".

The third law defines a simple relationship between a planet's period of revolution and its mean distance from the Sun: taking the year as the unit of time and the astronomical unit as the unit of distance, $P^2 = a^3$ where P is the period and a is the "mean distance".

The size and shape of an ellipse are determined by two quantities: the length of the "semi-major axis" and the "eccentricity". The semi-major axis is half the diameter of the ellipse measured at its widest points, and is equivalent to the "mean distance" used in Kepler's third law. The eccentricity is a measure of the ellipse's elongation: the larger the eccentricity, the "flatter" the ellipse. The value of the eccentricity will lie between 0 (for a circle) and 1 (for a straight line). The eccentricity of the Earth's orbit, for example, is relatively small, 0.0167, and its distance from the Sun ranges from $1.471 \times 10^8$ km at "perihelion" to $1.521 \times 10^8$ km at "aphelion".

The mean distances of the planets from the Sun range from 0.39 A.U. for Mercury to 39.4 A.U. for Pluto. Whereas most of the planetary orbits differ only slightly from circles, Mercury and Pluto are exceptional, having orbital eccentricities of 0.206 and 0.250 respectively; in the case of Pluto, its distance from the Sun varies from $4.443 \times 10^9$ km to $7.375 \times 10^9$ km. The orbital periods of the planets range from 87.97 days (Mercury) to 247.7 years (Pluto).

The apparent size and brightness of the Sun as seen from the various planets differ considerably. In the Mercurian sky the Sun would appear about 2.5 times larger and 6.3 times brighter than it does from the Earth, while at the mean distance of Pluto the Sun would appear only about one fortieth of the size and one sixteen-hundredth of the brightness seen from Earth.

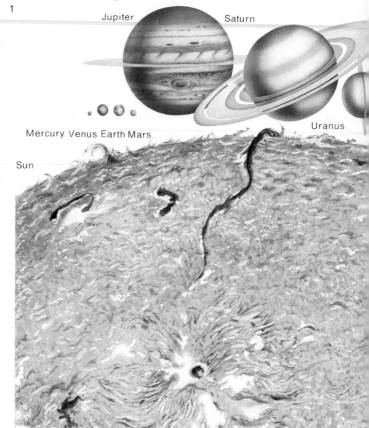

1 Jupiter Saturn / Mercury Venus Earth Mars / Uranus / Sun

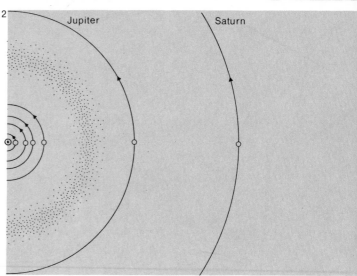

2 Jupiter Saturn

**Physical data for the Sun, Moon and planets**

| Name | Diameter | | Mass | Density | Volume | Surface gravity |
|---|---|---|---|---|---|---|
| | equatorial km | polar km | kg | (water = 1) | (Earth = 1) | (Earth = 1) |
| Sun | 1,392,530 | 1,392,530 | $1.9891 \times 10^{30}$ | 1.41 | $1.3 \times 10^6$ | 28.0 |
| Moon | 3,476 | 3,476 | $7.3483 \times 10^{22}$ | 3.34 | 0.02 | 0.165 |
| Mercury | 4,878 | 4,878 | $3.3022 \times 10^{23}$ | 5.43 | 0.06 | 0.377 |
| Venus | 12,104 | 12,104 | $4.8689 \times 10^{24}$ | 5.24 | 0.86 | 0.902 |
| Earth | 12,756 | 12,714 | $5.9742 \times 10^{24}$ | 5.52 | 1.00 | 1.000 |
| Mars | 6,794 | 6,759 | $6.4191 \times 10^{23}$ | 3.93 | 0.15 | 0.379 |
| Jupiter | 142,800 | 134,200 | $1.899 \times 10^{27}$ | 1.32 | 1,323 | 2.69 |
| Saturn | 120,000 | 108,000 | $5.684 \times 10^{26}$ | 0.70 | 752 | 1.19 |
| Uranus | 52,000 | 49,000 | $8.6978 \times 10^{25}$ | 1.25 | 64 | 0.93 |
| Neptune | 48,400 | 47,400 | $1.028 \times 10^{26}$ | 1.77 | 54 | 1.22 |
| Pluto | 3,000 | 3,000 | $6.6 \times 10^{23}$ | 4.7 | 0.01 | 0.20 |

# 1. Scale of the Sun and planets

The Sun has a radius of 696,265 km, nearly 10 times the radius of Jupiter, the largest of the planets, and over 100 times that of the Earth.

# 2. The Solar System

Nine major planets and many minor bodies orbit the Sun. The asteroid belt, dividing the inner planets from the outer planets, is shown as a speckled region. Pluto now lies within the orbit of Neptune, where it will remain for the next 20 years.

# 3. Planetary configurations

An "inferior" planet is one whose orbit lies inside that of the Earth; the others are said to be "superior". A superior planet is in "opposition" ($J_1$) when it lies directly opposite the Sun in the sky; it is in "quadrature" when the angle measured at the Earth between it and the Sun (called the "elongation") is 90° ($J_2$ and $J_4$). When either a superior or an inferior planet is directly in line with the Sun ($V_1$, $V_3$ and $J_3$) it is said to be in "conjunction". In the case of an inferior planet the conjunction may be either inferior ($V_1$) or superior ($V_3$); the maximum elongation of an inferior planet is shown at $V_2$.

3

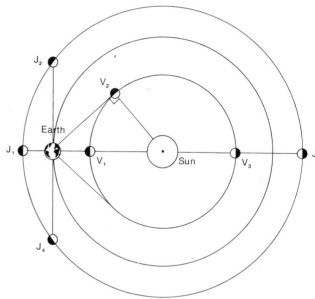

# 4. Kepler's Laws

According to Kepler's first law, the planets follow elliptical orbits around the Sun, with the Sun at one focus of the ellipse (**A**). An ellipse is an oval curve in which the sum of the distances from any point on the curve to two fixed points or "foci" is constant. Kepler's second law states that the radius vector of a planet sweeps out equal areas in equal times about the focus containing the Sun (**B**). The third law states that, for any planet, the square of its period of revolution is directly proportional to the cube of its mean distance from the Sun (**C**). By "mean distance" what is meant is the semi-major axis of the ellipse, i.e., half of its greatest diameter. Given the mean distance of any orbiting body it is simple to calculate its period, or vice versa.

4A

B

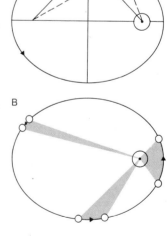

C

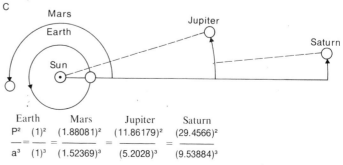

| | Earth | Mars | Jupiter | Saturn |
|---|---|---|---|---|
| $\dfrac{P^2}{a^3} =$ | $\dfrac{(1)^2}{(1)^3} =$ | $\dfrac{(1.88081)^2}{(1.52369)^3} =$ | $\dfrac{(11.86179)^2}{(5.2028)^3} =$ | $\dfrac{(29.4566)^2}{(9.53884)^3}$ |

●Pluto
eptune

Uranus · Pluto

Neptune

| Mean vis. opposition Mag. | Mean distance A.U. | millions of km | Eccentricity e | Inclination degrees | Sidereal period days | Albedo |
|---|---|---|---|---|---|---|
| −26.8 | — | — | — | 7.25 | — | — |
| −12.7 | — | — | — | 1.53 | — | 0.07 |
| 0.0 | 0.3870987 | 57.91 | 0.2056306 | 0 | 87.969 | 0.06 |
| −4.4 | 0.7233322 | 108.21 | 0.0067826 | 178 | 224.701 | 0.76 |
| — | 1.0000000 | 149.60 | 0.0167175 | 23.44 | 365.256 | 0.36 |
| −2.0 | 1.5236915 | 227.94 | 0.0933865 | 25.20 | 686.980 | 0.16 |
| −2.6 | 5.2028039 | 778.34 | 0.0484681 | 3.12 | 4332.59 | 0.73 |
| +0.7 | 9.5388437 | 1,427.01 | 0.0556125 | 26.73 | 10,759.20 | 0.76 |
| +5.5 | 19.181826 | 2,869.6 | 0.0472639 | 97.86 | 30,684.8 | 0.93 |
| +7.8 | 30.058021 | 4,496.7 | 0.0085904 | 29.56 | 60,190.5 | 0.62 |
| +14.9 | 39.44 | 5,900.0 | 0.250 | 90 | 90,465.0 | 0.5 |

# Time and the Seasons

It is natural that very early in man's history, the behavior of the Sun in the sky, dividing day from night and winter from summer, should have become important as a means of measuring time. From an Earth based or "geocentric" point of view, the Sun rises daily on the eastern horizon, reaching its highest point in the sky at noon, and finally sets behind the western horizon, apparently revolving around the Earth in much the same way as the stars appear to do. If observations are continued throughout the course of a year, it will be noticed that the Sun's maximum height above the horizon varies, and so does the length of time it remains above the horizon. These and other changes can be followed most clearly by reference to the "celestial sphere", an imaginary sphere of very large radius with the Earth at its center and with the stars considered as fixed points on its surface. The intersection of the celestial sphere with the plane defined by the Earth's equator marks the celestial equator, while the north and south celestial poles are located perpendicularly above the terrestrial poles.

As the Earth travels around the Sun, so the Sun appears from Earth to change its position relative to the background of stars: if the Sun's position on the celestial sphere is plotted daily, it is found to move eastwards with respect to the stars through an angle of approximately 1° per day. In the course of a year it traces out a complete circle on the celestial sphere. This circle, the apparent annual path of the Sun, is called the "ecliptic".

The ecliptic, by definition, lies in the same plane as the Earth's orbit; and since the Earth's equator is tilted to the orbital plane, so the ecliptic is tilted by the same angle (23° 27′) to the celestial equator. On or around 21 March each year the Sun crosses the celestial equator, going from south to north, and the point at which it crosses—one of the two points of intersection between the ecliptic and the celestial equator—is called the "vernal equinox". As the year continues, the Sun's angular distance from the celestial equator increases. (The angle, measured north or south, between the celestial equator and a celestial body is called the "declination" of the body, and is used as a coordinate roughly equivalent to latitude on the Earth.) After about three months (around 21 June) the Sun reaches its maximum northerly declination, a point known as the "summer solstice". Its declination then begins to decrease, and some three months later (around 23 September) the Sun recrosses the equator at a point directly opposite the vernal equinox called the "autumnal equinox". By about 21 December the Sun reaches its point of greatest southerly declination—the "winter solstice"—after which time it begins the final quarter of its annual path along the ecliptic.

## The seasons

The phenomenon of the seasons is a consequence of the fact that the Earth's axis of rotation is not perpendicular to the plane of its orbit, but tilted from the vertical by an angle of 23° 27′. Summer in the northern hemisphere occurs when the position of the Earth in its orbit is such that the north polar regions are tilted towards the Sun, while winter occurs six months later, when the north pole is tilted away from the Sun. In geocentric terms, these times correspond to the Sun's maximum (most northerly) and minimum (most southerly) declination as it moves along the ecliptic.

When the Sun lies at one of the equinoxes, the Sun is overhead at the equator, and the days and nights are of equal duration all over the globe. The solstices, on the other hand, mark the points of midsummer and midwinter, the longest and shortest days in the year respectively. (In the southern hemisphere, the seasons occur at the opposite times of the year, as do the longest and shortest days.)

For observers located within the "Arctic Circle" (a circle of radius 23° 27′ centered on the north pole) the Sun remains above the horizon for 24 hr per day when it is at the summer solstice. At this time of the year these observers experience the phenomenon of the "Midnight Sun". Six months later, however, the northern

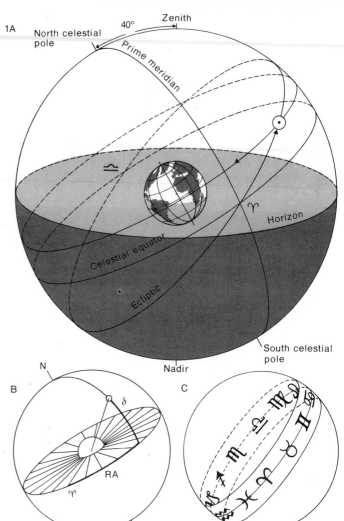

## 1. Celestial sphere

The intersection between the celestial equator and the ecliptic on the celestial sphere (**A**) defines the vernal equinox (♈). The Sun is shown at a point roughly midway between the vernal equinox and the summer solstice. Its apparent daily motion across the Earth's sky from east to west is also indicated. A simplified diagram (**B**) shows how the celestial coordinates of a body are defined: right ascension (R.A.) is measured eastward along the equator from the vernal equinox in units of hours, minutes and seconds; declination ($\delta$) is measured in degrees along a great circle passing through the celestial poles. The 12 signs of the Zodiac (**C**) lie in a band of about 9° on either side of the ecliptic. (These examples have been drawn for latitude 50° N: the angle between the north celestial pole and the observer's zenith is therefore 40°.)

## 2. Seasons

The Earth's rotational axis is inclined to the plane of its orbit. In the course of an orbit around the Sun, therefore, the Earth's

north and south hemispheres are alternately pointed slightly towards or away from the Sun. At midsummer in the northern hemisphere, for example, the Earth's north pole is inclined in the direction of the Sun; the days are consequently longer and the maximum altitude of the Sun above the horizon (at midday) is greater than at any other time during the year. Conversely, in midwinter the north pole leans away from the Sun, and the conditions are reversed. In the southern hemisphere the seasons occur at opposite times.

### Conversion of Mean Solar Time into Sidereal Time

| | |
|---|---|
| $24^h$ MST $\equiv (24^h + 3^m 56{}^s.556)$ ST |
| $1^h$ MST $\equiv ( 1^h + 9{}^s.8565)$ ST |
| $1^m$ MST $\equiv ( 1^m + 0{}^s.1643)$ ST |
| $1^s$ MST $\equiv ( 1^s + 0{}^s.0027)$ ST |

### Conversion of Sidereal Time into Mean Solar Time

| | |
|---|---|
| $24^h$ ST $\equiv (24^h - 3^m 55{}^s.910)$ MST |
| $1^h$ ST $\equiv ( 1^h - 9{}^s.8296)$ MSR |
| $1^m$ ST $\equiv ( 1^m - 0{}^s.1638)$ MST |
| $1^s$ ST $\equiv ( 1^s - 0{}^s.0027)$ MST |

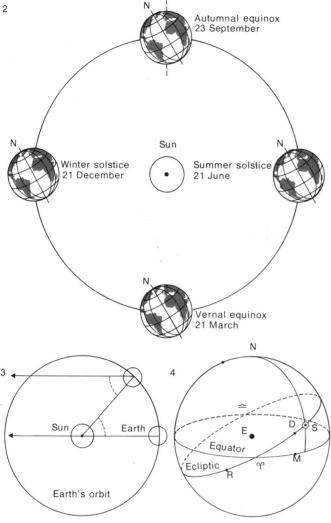

2

Autumnal equinox
23 September

Sun

Winter solstice
21 December

Summer solstice
21 June

Vernal equinox
21 March

3

Sun    Earth

Earth's orbit

4

N

E

D ⊙ S

Equator

M

Ecliptic    ♈

R

### 3. Sidereal time
The sidereal day is slightly shorter than the solar day because in the time it takes the Earth to complete one rotation relative to the fixed stars, the Earth will also have travelled about 1° along its orbit: it will have to rotate slightly further before the same observer is again directly facing the Sun.

### 4. Mean solar time
A fictitious body, the "dynamical mean sun" (D), is imagined moving along the ecliptic at the

Sun's mean rate, starting from perihelion (R) at the same moment as the real Sun. Mean solar time is defined by the motion of a second fictitious body, the "mean sun" (M), which sets off along the equator at the moment the dynamical mean sun reaches the vernal equinox, and moves at the same uniform rate.

### 5. Equation of time
The difference between mean time and apparent solar time varies through the course of a year.

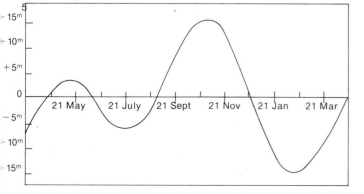

regions are shrouded in darkness, and the Sun never rises. Similar effects occur in the Antarctic Circle, but at opposite times.

**The Sun and time**
The apparent solar day is the time interval between two successive noons—noon being the instant at which the Sun is at its highest point above the horizon. This interval is divided into 24 hr of apparent solar time. However, the length of the apparent solar day varies by a small but significant amount during the course of a year. The reason is twofold. First, because the Earth's orbit is elliptical, the velocity of the Earth fluctuates in a periodic way, as the Earth moves faster when it is closer to the Sun than when it is further away (in accordance with Kepler's second law—*see* page 8). Consequently, the apparent motion of the Sun on the ecliptic varies correspondingly. Secondly, even if the Sun were to move along the ecliptic at a uniform rate, its apparent annual motion in relation to the celestial equator would vary because the ecliptic is inclined to the celestial equator. In other words, the Sun's "right ascension" increases at a non-uniform rate. ("Right ascension" is a coordinate on the celestial sphere, roughly equivalent to longitude on the Earth, and is measured eastward along the celestial equator from the vernal equinox; it is given in units of hours, minutes and seconds, such that 24 hr is equivalent to 360°.) Since it is the position of the Sun with respect to the equator that matters in determining the instant of noon, it follows that the length of the apparent solar day will vary in the course of a year.

In order to provide a system of timekeeping in which the units of time remain constant, "mean solar time" (or "mean time") is used. Mean time is based on a hypothetical point, the "mean Sun", which is imagined as moving along the celestial equator at a rate equal to the mean motion of the real Sun on the ecliptic: the mean Sun moves through an angle of 0°.0986 per day (360° divided by 365.26 days). The time interval between two successive mean noons is constant, and defines the length of the mean solar day, which is in turn divided into 24 hr of mean time.

The difference between apparent solar time and mean time is referred to as the "equation of time", a quantity that varies during the course of a year from a minimum value of about − 14.3 min to a maximum of about + 16.3 min. When the value of the equation of time is positive, apparent time is ahead of mean time.

In addition to mean solar time there is another important system of timekeeping, "sidereal time", which is independent of the Earth's orbit around the Sun, being based on its rotation about its own axis measured against the background of fixed stars. Although the rising and setting of the Sun is governed by the rotation of the Earth, the length of the solar day is not equal to the true axial rotation period of the Earth. As a result of the Earth's motion around the Sun the solar day is slightly longer than the sidereal day: in the time it takes the Earth to rotate once about its axis the Earth also moves a small amount (about 1°) along the path of its orbit. An observer who faces directly towards the Sun at the start of one rotation will have to wait for slightly longer than a complete rotation of the Earth before he is once again facing the Sun (*see* diagram 3). One sidereal day is equal to 23 hr 56 min 4.1 sec of mean solar time. In the course of a year there are about 365 complete solar days, but the Earth actually makes about 366 full rotations on its axis; the motion of the Earth around the Sun in effect cancels out one complete axial rotation relative to the Sun.

The sidereal day is equivalent to one rotation of the celestial sphere. Like the mean solar day, it is divided into 24 hr of sidereal time, which is in turn divided into sidereal minutes and sidereal seconds. For a given point on the Earth's surface, the sidereal day is measured from the moment at which the vernal equinox crosses the observer's meridian (the great circle on the celestial sphere running through the celestial poles and the observer's zenith—*see* Glossary). This moment defines 0 hr of local sidereal time.

5
− 15ᵐ
− 10ᵐ
+ 5ᵐ
0
− 5ᵐ
− 10ᵐ
− 15ᵐ

21 May    21 July    21 Sept    21 Nov    21 Jan    21 Mar

# Eclipses

The Earth has one natural satellite, the Moon, which by chance appears almost exactly the same size in the sky as the Sun: for although the Sun is nearly 400 times larger in diameter than the Moon, it is also almost 400 times further away. When the Earth passes into the long shadow cast into space by the Moon, an eclipse of the Sun occurs; similarly, when the Moon passes into the Earth's shadow there is an eclipse of the Moon.

If the lunar orbit lay in the plane of the Earth's orbit (the plane of the ecliptic) there would be a solar eclipse every new Moon and a lunar eclipse every full Moon. However, the lunar orbit is inclined to the ecliptic by an angle of 5° 9′ (a value which fluctuates by a small amount) and consequently, on most occasions, the shadow of the new Moon passes above or below the Earth, while at full Moon the Moon usually passes above or below the Earth's shadow. For an eclipse to take place, the Moon must be close to one of its "nodes", the points on its orbit at which it crosses the ecliptic.

The lunar shadow is divided into two regions, the dark, cone-shaped "umbra", and the surrounding "penumbra". From within the umbra the Sun's disc is completely obscured, so that an observer in this region would see a "total" eclipse; an observer within the penumbra would see a "partial" eclipse, with only part of the solar disc hidden.

Because the Moon's orbit is elliptical, its distance from the Earth's center varies. The closest approach (363,300 km) is called "perigee" and the most distant point (405,600 km) is called "apogee". Sometimes a solar eclipse occurs at a time when the Moon is relatively distant from the Earth, and the Moon's umbra does not stretch far enough to reach the Earth's surface. On such occasions the Moon does not appear sufficiently large to obscure the entire solar disc, and the black disc of the Moon is seen surrounded by a ring of light; this type of eclipse is called "annular". During a lunar eclipse, the Moon does not usually vanish completely, because some of the Sun's rays are bent or "refracted" onto the lunar surface by the Earth's atmosphere. This effect sometimes gives the Moon a reddish appearance during eclipse.

## Duration and frequency

In a solar eclipse the shadow of the Moon crosses the Earth at a speed of about 3,000 km hr$^{-1}$ and the umbra traces a track over part of the terrestrial surface. The maximum possible duration of totality at a given place is 7 min 31 sec. The longest eclipse in recent times took place on 20 June 1955 and lasted for 7 min 8 sec; the next totality of long duration will be of 6 min 54 sec on 11 July 1991. Although total eclipses are not particularly rare events, a given point on the Earth's surface will experience one only on very infrequent occasions. The last total eclipse to be seen from anywhere in England, for example, took place on 29 June 1927, and the next will occur on 11 August 1999.

From a given location on Earth more eclipses of the Moon than of the Sun will be seen, but, in fact, more eclipses take place of the Sun than of the Moon. The reason for this is that any lunar eclipse may be seen simultaneously from a complete hemisphere of the Earth while a solar eclipse is seen only from that part of the terrestrial surface on which the shadow falls. The maximum possible number of eclipses of the Sun and Moon in any one year is seven, of which four or five will be solar eclipses; the minimum number of eclipses that can occur in a year is two, both of which will be solar eclipses.

## Eclipse cycles

The ancient Babylonians discovered that closely similar solar and lunar eclipses recur at intervals of 18 years 10 or 11 days (depending on the number of "leap years" in the period). This eclipse cycle, known as the Saros, arises because of a curious relationship between the length of the synodic lunar month (the interval between one new Moon and the next) and the motion of the Moon's

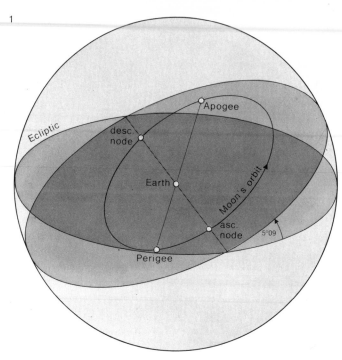

**1. The Moon's orbit**
The plane of the Moon's orbit is inclined to the ecliptic by an angle of 5° 9′. The Moon's nodes move backwards along the ecliptic at a rate of just over 19° per year, while the line of apsides (joining perigee and apogee) moves in the opposite direction with a period of 8.85 years.

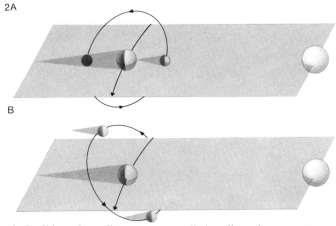

**2. Conditions of an eclipse**
An eclipse, either lunar or solar, can only occur if there is a new or full Moon when the Moon is near one of its nodes (**A**). Since the Moon's orbit is inclined to the ecliptic, eclipses do not occur every lunation; instead the shadow of the Moon may pass either over or under the Earth, while that of the Earth may pass over or under the Moon (**B**).

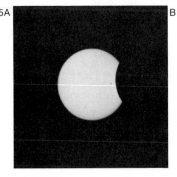

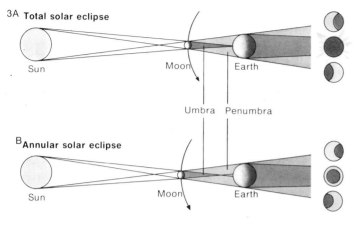

**3A Total solar eclipse**

Sun     Moon   Earth

Umbra   Penumbra

**B Annular solar eclipse**

Sun     Moon   Earth

## 3. Types of eclipse
The Moon's shadow is divided into two regions, the dark central "umbra" and the lighter "penumbra", within which part of the Sun remains visible. A total eclipse of the Sun occurs when the Earth passes directly into the shadow cast by the Moon (**A**). However, the eclipse only appears total from the limited region of the Earth's surface that is covered by the umbra; from inside the penumbra the eclipse is partial. An annular eclipse (**B**) occurs when the Moon is near apogee,

and its shadow cone does not reach the Earth. The angular size of the Moon as seen from Earth is therefore too small to cover the Sun's disc, so that a thin ring of light remains visible around the black disc of the Moon.

## 4. Eclipse tracks
The tracks of total solar eclipses as calculated for the period 1981–2001 have been plotted onto this map of the Earth. Only 14 total eclipses will occur. Each track represents a duration less than 8 min.

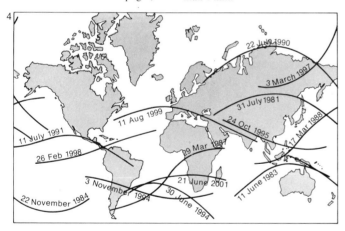

## 5. Eclipse sequence
A total eclipse (**A**) begins with the Moon's disc encroaching on the west side of the Sun, steadily obscuring a larger and larger area (**B**). After about an hour totality

is reached (**C**). The intensity of light drops to the level of twilight, and the corona and chromosphere become visible. Lastly the photosphere reappears, producing a "diamond ring" effect (**D**).

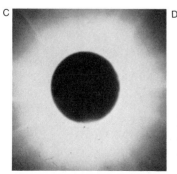

C

D

orbit. The Sun passes through a given node of the lunar orbit at intervals of 346.62 mean solar days, a period of time known as the "eclipse year". By coincidence, 19 eclipse years (6,585.78 mean solar days) is very nearly equal to 223 synodic months (6,585.32 mean solar days). Consequently almost identical eclipse conditions will recur at intervals of 223 synodic months.

By a further coincidence, if the interval between the Moon's closest approaches to the Earth (27.55455 days) is multiplied by 239, the result is 6,585.54 days—again very close to the Saros period. Thus if, for example, a solar eclipse were to occur on one occasion with the Moon at perigee, the next eclipse in the Saros would occur 6,585.32 days later and the Moon, once again, would be very close to perigee. If it were not for the fact that the motion of perigee fits in so neatly with the Saros period, successive eclipses of the cycle would differ widely in duration and nature. Instead, the complementary nature of the two factors ensures that two successive eclipses in a Saros are closely similar. For example, in the eclipse of 30 June 1973 totality lasted for a maximum of 7 min 04 sec, while the duration of totality in the eclipse that is due to occur after one Saros period—on 11 July 1991—should be about 6 min 54 sec.

Because the various periods involved are not precisely equal, the repetition of eclipses in a Saros does not continue indefinitely. A complete Saros lasts, typically, for more than 1,000 years, beginning with a series of partial eclipses visible from one of the polar regions and developing into a sequence of total (or annular) eclipses; the final eclipses of the Saros will again be partials, visible from the opposite polar region. Between successive eclipses of one particular Saros many other eclipses will occur, each belonging to different Saros cycles. New cycles begin from time to time, and old cycles come to an end. For example, the partial eclipse of 12 August 1942, visible in the Antarctic, was the last of one Saros, while the partial visible from the Arctic on 17 June 1928 marked the beginning of a new Saros.

The first recorded successful application of the Saros cycle to predict an eclipse is believed to have been achieved by the Greek philosopher Thales of Miletus in respect of the eclipse of 28 May 585 BC. The amateur astronomer today may easily use the Saros period to predict the date of some forthcoming eclipses by adding the appropriate interval to the dates of past eclipses. Calculation of the exact path of the eclipse is much more difficult, although a rough estimate of the general region of the Earth from which an eclipse may be seen can be made: the Saros period comprises a whole number of days plus 0.32 day. During one third of a day the Earth rotates through an angle of 120°; the eclipse track of a given eclipse will therefore lie roughly 120° to the west of the track of the previous eclipse in the cycle.

### Scientific importance of eclipses
Careful analysis of historical eclipse data shows that the times and locations of ancient eclipses are not entirely consistent with the present motion of the Moon and rotation of the Earth. By such means it has been shown that the tidal interaction between the Earth and Moon is pushing the Moon away from the Earth at a rate of about 4 cm per year, while the rotation period of the Earth is increasing (in other words, the day is lengthening) by about 0.002 sec per century.

Another application of total eclipses led to the dramatic confirmation of one of the predictions of Einstein's General Theory of Relativity, which is the best current theory of gravitation. According to this theory light rays should be deflected as they pass close to a massive body such as the Sun. A group from the Royal Observatory, Greenwich, led by Sir Arthur Eddington, photographed the positions of stars lying close to the edge of the Sun during the total eclipse of 29 May 1919 and found them to be displaced from their normal positions by an amount which fitted well with the predictions of the theory.

# Structure of the Sun

Looked at in cross-section, the Sun may be divided into a number of concentric layers, or shells, each of which will be treated in more detail in later pages. The visible surface is known as the "photosphere" (meaning "sphere of light"), but it is not a solid surface like that of the Earth; rather it is a shallow shell of gas, some 400 km thick, from which all but a tiny fraction of the Sun's light is emitted. Below the photosphere, where the temperature and density increase rapidly, the solar material is opaque to visible light, while above it the solar atmosphere is essentially transparent. The solar radius, measured from the center of the Sun to the photosphere, is 696,265 km; since most of the visible light emanates from a region within the photosphere only about 100 km thick, it is hardly surprising that the edge of the solar disc appears hard and sharp.

The Sun shines as a result of nuclear reactions, releasing vast quantities of energy, which take place in the hot, dense central core, a region extending to about 0.25 of the solar radius, a distance of some 175,000 km from the center (this value is not known with certainty). The central temperature is about 15 million K and the density is estimated to be 160,000 kg m$^{-3}$. Material outside the core is at lower temperatures, pressures and densities, and nuclear reactions do not take place at significant rates.

Energy is released in the core primarily in the form of $\gamma$-ray and X-ray radiation or high-energy, short wavelength "photons" (*see* pages 82–83). These photons cannot travel far before being absorbed: deep in the solar interior the mean distance travelled is about 1 cm. The photons are absorbed and reemitted a multitude of times, and their energies decline as they move out through the solar globe. Their paths take the form of a "random walk", in which the direction and energy change at each absorption and emission. As a result, it can take something like $10^4$ to $10^5$ years for a given packet of energy to make its way to the surface. This process of absorption, scattering and emission goes on through a region called the "radiative zone", which extends from about 0.25 solar radii ($R_S$) to over 0.8 $R_S$ (a value of 0.86 $R_S$ is commonly quoted). Photons, absorbed and emitted many times, are progressively converted from $\gamma$-ray and X-ray radiation into extreme ultraviolet and ultraviolet, finally emerging from the photosphere as visible light (low-energy or longish wavelength photons).

Beyond about 0.8 $R_S$ the lower temperature allows electrons to be captured by nuclei—particularly those of heavier elements—to form partially ionized atoms, which are very effective at capturing photons. As a result, the solar material becomes very much more opaque than in the layers below (by a factor of about 20) and there is much greater resistance to the flow of radiation out through the Sun. A steep temperature gradient is established, and hot bubbles of gas rising through this gradient are accelerated to form convective cells. Below about 0.85 $R_S$ convection is not believed to be significant, and the transport of energy from the interior is mainly by radiation. Above this level, however, where blocking of radiative transfer takes place, convection sets in and masses of hot gas are carried bodily to the surface, where they radiate their heat. In the outermost 15 to 20 percent of the solar radius, energy transport is primarily by convection; this layer is known as the "convective zone".

At the top of the convective zone lies the photosphere. Under ideal observing conditions, the photosphere may be seen to be made up of a large number of bright "granules" (*see* pages 30–31). These granules are often polygonal in shape, and are separated by thin dark lines. They are evidence of the bubbling motion of hot gases in the outer parts of the solar globe in the process of transporting heat outward from the interior by convection. On a much larger scale there is a network of "supergranular" cells, each of which contains hundreds of individual granules. Being larger patterns of circulation, they extend deeper into the convective zone.

The photosphere is overlaid by the "chromosphere", a more rarefied layer a few thousand kilometers thick. The temperature at

the base of the chromosphere is about 4,300 K, but it increases with altitude, rising very sharply through a thin transition region, where it merges with the "corona", an even more rarefied region, having a density at base of about $10^{-11}\,kg\,m^{-3}$ and extending outward to a distance of several solar radii. The height of the transition region above the photosphere varies quite markedly over the solar surface, but is typically about 2,000 km.

The corona has a very high temperature of between 1 million and 5 million K but, because its density is so low, the amount of heat it contains is very small compared with the photosphere. Indeed, only about one millionth of the Sun's visible light comes from the corona, and even this is photospheric light scattered in the corona rather than light emitted by the corona. Under normal circumstances the chromosphere and corona cannot be seen without the aid of specialized instrumentation, but when a total eclipse occurs these layers may be seen directly. Beyond the corona there extends the solar "magnetosphere" and the "solar wind", the stream of atomic particles (mainly electrons and protons) which flows away from the Sun into interplanetary space (*see* pages 76–77).

The above is a summary of the structural properties of the quiet Sun. However, the Sun is an active star. Violent events occur on it, the most dramatic being explosive flares, while "prominences" (huge flame-like clouds of luminous gas) may be observed above the chromosphere. Aspects of solar activity and variability will be treated in later pages.

**Physical characteristics of the Sun**

| | |
|---|---|
| Diameter | 1,392,530 km |
| Volume | $1.41 \times 10^{18}\,m^3$ |
| Mass | $1.9891 \times 10^{30}\,kg$ |
| Magnetic field strengths (typical): | |
| Sunspots | 3,000 G |
| Polar field | 1 G |
| Bright, chromospheric network | 25 G |
| Emphemeral (unipolar) active regions | 20 G |
| Chromospheric plages | 200 G |
| Prominences | 10 to 100 G |

Chemical composition of photosphere

| Element | % weight | Element | % weight |
|---|---|---|---|
| Hydrogen | 73.46 | Nitrogen | .09 |
| Helium | 24.85 | Silicon | .07 |
| Oxygen | .77 | Magnesium | .05 |
| Carbon | .29 | Sulphur | .04 |
| Iron | .16 | Other | .10 |
| Neon | .12 | | |

| | |
|---|---|
| Density (water = 1): | |
| Mean density of entire Sun | $1,410\,kg\,m^{-3}$ |
| Interior (center of Sun) | $1.6 \times 10^5\,kg\,m^{-3}$ |
| Surface (photosphere) | $10^{-6}\,kg\,m^{-3}$ |
| Chromosphere | $10^{-9}\,kg\,m^{-3}$ |
| Low corona | $10^{-13}\,kg\,m^{-3}$ |
| Solar radiation: | |
| Entire Sun | $3.83 \times 10^{23}\,kW$ |
| Unit area of surface of Sun | $6.29 \times 10^4\,kW\,m^{-2}$ |
| Received at top of Earth's atmosphere | $1,368\,W\,m^{-2}$ |
| Surface brightness of the Sun (photosphere): | |
| Compared to full Moon | 398,000 times |
| Compared to inner corona | 300,000 times |
| Compared to outer corona | $10^{10}$ times |
| Temperature: | |
| Interior (center) | 15,000,000 K |
| Surface (photosphere) | 6,050 K |
| Sunspot umbra (typical) | 4,240 K |
| Penumbra (typical) | 5,680 K |
| Chromosphere | 4,300 to 50,000 K |
| Corona | 800,000 to 3,000,000 K |
| | |
| Rotation (as seen from Earth): | |
| Of solar equator | 26.8 days |
| At solar latitude 30° | 28.2 days |
| At solar latitude 60° | 30.8 days |
| At solar latitude 75° | 31.8 days |

# History I

From earliest times man must have studied the Sun and its movements in the heavens with a sense of awe and wonder. As the provider of heat and light the Sun was, and is, vital to life on Earth; it is hardly surprising that a belief in the divinity of the Sun was strongly held in early civilizations. The seasonal changes in the position and motion of the Sun were well known by the dawn of recorded history, and on the basis of these movements the agricultural peoples of Egypt had established a calendar based on a year of 365 days by about 3000 BC. Eclipses must have attracted fascinated attention, the earliest known record of such an event dating back to the eclipse of 2136 BC seen in China. In the Middle East, the Babylonians and Chaldeans had begun the regular tabulation of eclipses by 747 BC, while the earliest occasion on which an eclipse is believed to have been predicted was in 585 BC (*see* page 13).

### Location of the Sun: geocentric and heliocentric theories
All early civilizations believed that the Earth lay at the center of the universe. The geocentric view was developed to its fullest extent by the Greeks, who conceived a system in which the Sun, the Moon, the stars and planets revolved around the Earth in perfect circular motions. It was apparent to them that the observed motions could not be explained simply in terms of uniform motion on a circle centered on the Earth, and elaborate schemes involving combinations of circular motions were devised to obtain better agreement between theory and observation. Best known of these devices was the epicycle, a circle whose center travels round the circumference of another circle. The Greek view of the universe was synthesized by Claudius Ptolemaeus (Ptolemy) in his book the *Almagest*, published about the middle of the second century AD. This view remained largely unchallenged for over a millenium.

Long before the time of Ptolemy, a few philosophers had opposed the geocentric theory; most notably, Aristarchus (310–250 BC) had suggested that the Sun was located at the center of the universe and that the Earth and planets travelled around it. Many objections were raised against this idea, which was contrary to the accepted laws of nature of the time, and which also offended against prevailing religious views, and it was not until the fifteenth century AD that this "heliocentric" ("sun-centered") theory was raised again as a serious proposition.

In 1543, the Polish astronomer Nicolaus Copernicus (1473–1543) published a detailed account, *De Revolutionibus Orbium Coelestium*, of his heliocentric theory which, despite considerable opposition, came to be espoused by many leading thinkers, including the great Italian natural philosopher Galileo Galilei (1564–1642). Because Copernicus had retained the notion of uniform circular motion, his heliocentric theory was still unable fully to explain the observed motions of celestial bodies. In 1609, however, the German astronomer Johannes Kepler (1571–1630) demonstrated that the orbits of the planets around the Sun were elliptical; his three laws of planetary motion (*see* page 8) accounted fully for the motions of the Sun, the Moon and the planets, and dealt the final blow to the old geocentric theory.

The idea that the Sun might be a star like any other also began to emerge at about the same time—although it has been suggested that this view had been proposed by Aristarchus.

### Distance and size of the Sun
In one of the earliest attempts to describe the physical nature of the Sun, the Greek philosopher Anaxagoras (*c.*499–*c.*427 BC) argued that the Sun was a mass of fiery hot stone about the size of the Peloponnesus (the southern peninsula of Greece) and was banished from Athens for having the temerity to make such a suggestion. The determination of the size of the Sun depends upon an accurate knowledge of its distance; given the distance, a precise measurement of the apparent angular size of the Sun will allow its linear diameter to be calculated.

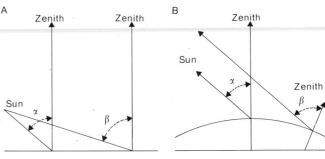

**1. Anaxagoras' estimate of the distance of the Sun**
On the assumption that the Earth was flat, Anaxagoras probably measured the angles $\alpha$ and $\beta$ from two places a known distance apart and calculated what he took to be the distance of the Sun by triangulation (**A**). In reality most of the difference in angles results from the Earth's curvature (**B**).

**2. Aristarchus' method**
Aristarchus assumed that the Moon travelled at a uniform speed in a perfect circle around the Earth. By measuring the difference in time between the first and third quarter and between the third and first quarter, he deduced angular values from which he calculated that the Sun was 19 times further away than the Moon.

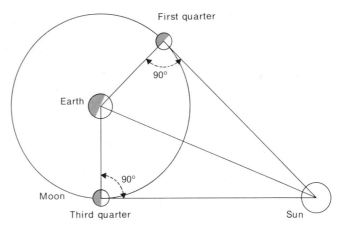

**3. Early cosmological systems**
In the Ptolemaic universe (**A**) the Earth was surrounded by a system of concentric spheres to which the planets and stars were fixed. The Copernican system (**B**) placed the Sun at the center of the universe. The Tychonic system (**C**) was a hybrid, with the Earth at the center but with the planets revolving round the Sun.

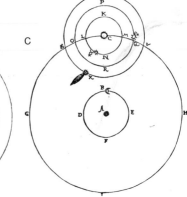

**4. Nicolaus Copernicus (1473–1543)**
Copernicus elaborated his heliocentric system of the universe in *De Revolutionibus Orbium Coelestium*, published in 1543.

**5. Tycho Brahe (1546–1601)**
Although Brahe's theoretical work had little permanent value, his detailed observations provided the basis for Kepler's laws.

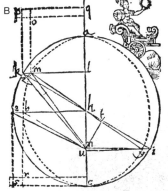

**6. Johannes Kepler (1571–1630)**
It was Kepler (**A**) who placed the heliocentric view of the Solar System on a secure basis. He published his key discovery that

the orbit of Mars was elliptical rather than circular (**B**) in *Astronomia Nova* (1609). Newton later established the dynamical basis for Kepler's laws.

**7. Transits of Venus**
As Venus passes between the Sun and the Earth (**A**), observers at different latitudes, X and Y, see Venus follow different tracks, XX¹ and YY¹ across the solar disc (**B**). By timing the instants at which Venus enters and leaves the solar disc (X, Y, and X¹, Y¹, respectively) the precise lengths of the lines XX¹ and YY¹ may be found. From this

information the angular separation between the two lines can be determined. Knowing, from Kepler's laws, the *relative* distances from the Sun of the Earth and of Venus, and having measured by this method the parallax of Venus from sites X and Y separated by a known distance, the distance of the Sun can be found from geometry.

In about 270 BC Aristarchus attempted to measure the distance of the Sun by an ingenious geometric technique (*see* diagram 2). The difficulty in measuring the appropriate angles led him to underestimate grossly the solar distance: he found the Sun to be about 19 times further away than the Moon, an underestimate by a factor of 20. Hipparchus (second century BC) used observations of the Earth's shadow on the Moon during a lunar eclipse to obtain a better estimate of the lunar distance. He then used his result, in conjunction with Aristarchus' work, to obtain an improved value of the Sun's distance which, in modern terms, amounted to about 10 million km. He was able to show that the Sun must be at least seven times the size of the Earth.

No substantial improvement in the value of the astronomical unit occurred until 1672, when the Italian astronomer G. D. Cassini used measurements of the parallax of the planet Mars to obtain a value of 138,370,000 km. The *relative* distances of the planets from the Sun could be established fairly readily (by means of, for instance, Kepler's third law—*see* page 8) so, it was only necessary to find the distance from the Earth to a planet in order to work out the Sun's distance.

The planet Venus, which approaches closer than Mars, moves in an orbit inclined to the ecliptic by an angle of 3° 24′ and, when at its closest, usually passes just above or just below the Sun in the sky. Occasionally, however, when inferior conjunction occurs with Venus close to one of the nodes of its orbit, it will pass directly between Earth and Sun and may be seen as a small black dot crossing the solar disc; it is then said to be in transit. Following an earlier suggestion by the Scottish astronomer James Gregory, Edmond Halley, in 1678, worked out in detail a method of finding the astronomical unit from observations of transits of Venus (*see* diagram 7).

Unfortunately, transits of Venus are rare events: transits since that time have occurred only in 1761, 1769, 1874 and 1882. The best result from the earlier pair, due to Leonhard Euler, was 151,225,000 km; from the latter pair, Sir George Airy obtained a value of 150,152,000 km.

Parallax measurements of asteroids, some of which approach the Earth considerably more closely than does Venus, have also been used; for example, the close approach of Eros in 1931 allowed Sir Harold Spencer Jones to compute an improved value of 149,654,000 km. Modern determinations are based on radar techniques. A radar beam was first successfully bounced back from the Sun in 1959 by a Stanford University team headed by V. R. Eshleman. However the Sun, having a diffuse atmosphere (the corona) which reflects radio waves, is not a very good radar target, and instead radar measurements of Venus and other planets are used to provide the most accurate value of the astronomical unit. These measurements, taken together with the laws of celestial mechanics, have yielded the current value of 149,597,892 km, which is probably accurate to about 1 km. The resultant value for the Sun's diameter is 1,392,530 km.

**Sunspots**
Although the Sun was long believed to be a pure unblemished orb, naked-eye observations of dark patches on the photosphere— "sunspots"—date back well over 2,000 years. Under good conditions, when the Sun is dimmed by haze or some other atmospheric phenomenon, spots larger than about 40,000 km in diameter may be seen without telescopic aid; such spots make up about 1 percent of the total. Well established Chinese records of such sightings date back to 28 BC (and possibly to 165 BC), but the dark patches were generally assumed to be foreground objects, such as birds, or one of the two inferior planets, Mercury and Venus. The belief in the flawless quality of the Sun was so great that even Kepler, in 1607, ascribed the sighting of a naked-eye sunspot to the transit of Mercury across the solar disc.

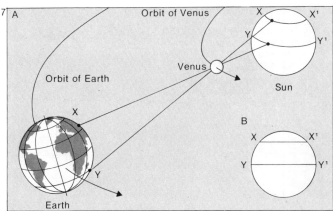

# History II

Following the invention of the telescope, usually (although perhaps erroneously) ascribed to the Dutch spectacle maker Hans Lippershey in 1608, a number of observers began to make telescopic solar observations. Galileo claimed to have observed sunspots in November 1610 (although he did not publish these results until 1612), and his subsequent observations led him to the conclusion that the Sun rotates on its axis in a period of about one month. He clearly was convinced that the spots were located on the solar surface and were carried around by the Sun's axial rotation.

The most comprehensive early study of sunspots was made by Christoph Scheiner (1575–1650), a German Jesuit who made a long series of observations between 1611 and 1625 and published these in 1630 in a work dedicated to the Duke of Orsini, *Rosa Ursini sive Sol*. He not only deduced a rotation period (as seen from the Earth) of 27 days but also, from observations of changes in the apparent track of spots across the disc during the course of the year, deduced that the tilt of the Sun's axis relative to the ecliptic lay between 6° and 8°; it was well over 200 years before this figure was improved upon. He noted the structure of larger sunspots and also detected the presence, close to sunspot groups near the edge of the solar disc, of bright patches, now known as faculae.

Careful observations carried out between 1826 and 1843 by Samuel Heinrich Schwabe of Dessau, Germany, revealed that the number of sunspots increases and decreases in a cyclic way over a period of roughly a decade (*see* pages 40–41).

One of the outstanding solar observers of the nineteenth century was the English amateur Richard Carrington. The results of nearly a decade's consistent observation, published in 1863, revealed that sunspots at different latitudes on the Sun rotate in different periods, showing that the Sun has "differential rotation": the equatorial regions rotate in a shorter time than regions of higher latitude (*see* pages 38–39). He improved the value of the inclination of the solar axis and established a mean solar rotation period of 25.38 days, a figure that is still in use today, together with his system of numbering solar rotations, beginning with rotation 1, which commenced on 9 November 1853. It was Carrington, too, who first observed what has come to be known as a solar flare (*see* pages 66–69). On 1 September 1859 he observed "two patches of intensely bright and white light" suddenly appear within a major sunspot group, then fade away within a few minutes. Only very rarely do these explosive events reveal themselves in ordinary white light. This same flare was also witnessed independently by another English astronomer, R. Hodgson.

The corona (*see* pages 70–75) is believed to have been mentioned first in the writings of Plutarch (AD 46–120), and there are definite records of its having been seen at the eclipse of 22 December 968. For some considerable time there was debate as to whether or not it was truly associated with the Sun rather than the Moon or the Earth's atmosphere. Prominences (*see* pages 46–48) are considered first to have been studied in detail by the Swedish observer Vassenius at the total eclipse of 1733; they were certainly recorded at the eclipse of 1842 although, again, for some years there was no proof that these phenomena were associated with the Sun rather than the Moon.

## The spectroscope

Crucial to the advancement of solar studies, and to astronomy as a whole, was the development of spectroscopy. In 1666 Isaac Newton passed a beam of sunlight through a glass prism and found that it was spread out into a rainbow band of color ranging from red to violet. The reason for this is that when light is passed through a glass prism (or a lens) it is bent or "refracted" through a certain angle, and red light is bent less than blue light. Newton's experiment showed that white light is a mixture of all the different colors that appeared in his rainbow or "spectrum".

Light may be pictured as a kind of wave motion travelling

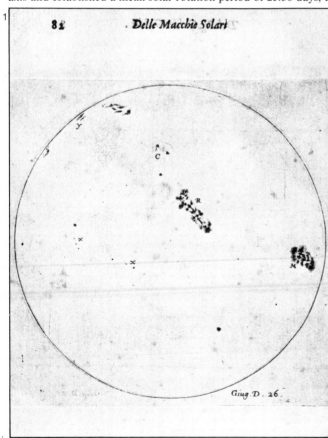

**1. Sunspot observations by Galileo**
The division of sunspots into umbral and penumbral regions is clearly apparent in this drawing from Galileo's *Istoria* (1613).

**2. Galileo Galilei (1564–1642)**
Galileo claimed to have been the first to discover sunspots, although he did not immediately recognize their significance. Two other astronomers (Fabricius and Scheiner) in fact published their observations first, but it was Galileo's *Letters on Sunspots* (1613) that brought sunspots to public attention.

**3. Christoph Scheiner (1575–1650)**
The frontispiece of Fr. Scheiner's *Rosa Ursina sive Sol* (1630) shows Scheiner making observations with the help of an assistant.

through space in something like the way waves travel in water. The distance between successive crests is the "wavelength", while the number of crests arriving at a given point per second is the "frequency" of the wave. Different wavelengths of light are perceived by the human eye as different colors: red light, for example, has a longer wavelength than blue. Thus another way of describing Newton's result would be to say that short wavelengths are refracted more than long wavelengths and that white light is a mixture of different wavelengths. Visible light is a very narrow portion of a much wider spectrum of radiation extending to much shorter and much longer wavelengths on either side (*see* page 26).

The spectroscope is an instrument for examining the spectrum. In essence it consists of a narrow slit through which light is admitted to a collimating system that passes the incoming light to a prism or series of prisms to split up the light. The resultant spectrum is either viewed by a telescope system or projected onto a photographic plate. Frequently a "diffraction grating" (a very finely ruled glass plate) is used as an alternative to the prism.

In 1802 the English chemist William Wollaston discovered that the continuous rainbow band of colors in the solar spectrum was crossed by a number of dark lines. These lines were subsequently studied in detail by the German physicist Joseph von Fraunhofer, who in 1814 mapped the wavelengths of 324 of them. The explanation of the presence of these lines did not come until 1859, when experimental work carried out by Gustav Kirchhoff, together with Robert Bunsen, led to the following laws of spectroscopy:
(1) An incandescent solid or a hot, dense (high pressure) gas will emit a continuous spectrum, i.e. a rainbow band of all colors;
(2) An incandescent gas under low pressure will emit an emission line spectrum, i.e. it will emit light only at a number of particular wavelengths, giving a spectrum consisting of a pattern of bright lines on a dark background;
(3) If a continuous spectrum is passed through a tenuous gas (under low pressure) that gas will absorb light at certain wavelengths (the same wavelengths at which it would emit in case (2) above) giving rise to a pattern of dark absorption lines on a bright background.

Essentially the continuous spectrum of the Sun consists of radiation generated in the hot dense interior which is emitted from the photosphere; the dark absorption lines are produced in the cooler, more rarefied region of the solar atmosphere above the base of the dense photosphere (this description is somewhat over-simplified—*see* pages 34–35).

It was shown that each chemical element produces its own characteristic pattern of lines so that, in principle, the chemical composition of the outer layers of the Sun could be deduced; modern physics accounts for the presence of these lines in terms of the physics of the atom (*see* pages 26–27). Detailed analysis of the spectrum can yield a tremendous range of information apart from chemical composition such as, for example, temperature, density, velocity, rotation, and the presence of magnetic fields. The importance of spectroscopy can scarcely be overrated.

**Photography**
The advent of photography offered a means of recording a solar image in a fraction of a second. The first good photograph of the Sun was obtained on 2 April 1845 by H. Fizeau and L. Foucault, in France, using the Daguerrotype process; in 1851 Berkowski made the first successful photograph of a total eclipse, revealing prominences and the corona.

A program of daily solar photography (weather permitting) was initiated, using a 8.9 cm photoheliograph built by Warren de la Rue, at the Kew Observatory, England, in 1858; from 1872 until 1980 this work was continued at the Royal Observatory. Since 1908, daily solar photography has been carried out using the 18 m tall tower telescope at Mount Wilson in California, and this process has continued at various centers throughout the world.

**4. Isaac Newton (1642–1727)**
Newton's work on optics led him to the discovery that white light from the Sun can be split into its component colors by means of a glass prism. The spectrum of colors produced in this way represents the range of wavelengths of which sunlight is composed, since the prism bends light of different wavelengths through different angles.

**5. Fraunhofer spectrum**
Absorption lines in the Sun's spectrum were studied by Joseph Fraunhofer in the early nineteenth century. Similar work had been carried out a few years previously by William Wollaston. The dark lines appear as a result of the absorption of certain wavelengths by atoms in the Sun's atmosphere (or, in some cases, in the Earth's atmosphere). Fraunhofer designated some of the most prominent lines by the letters A to I. The curve in this illustration represents the distribution of intensity with wavelength.

**6. Early photography**
The quality of photographs of the Sun taken in the nineteenth century is often remarkably high. This example was taken by L. M. Rutherfurd on 21 September 1870. There are numerous clearly defined sunspot groups, and several faculae (*see* page 31).

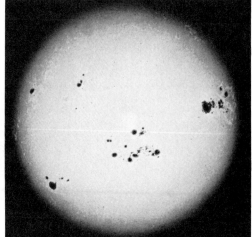

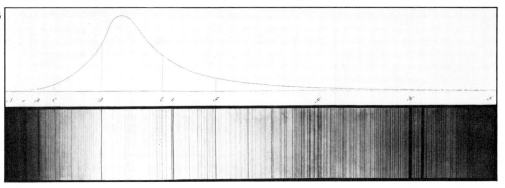

# History III

In 1868 the French astronomer Jules Janssen and the English observer Norman Lockyer independently discovered that prominences at the edge of the solar disc emitted light at particular wavelengths which could be detected by locating the slit of a spectroscope at a tangent to the edge of the Sun; in this way the brilliant light from the photosphere did not enter the instrument. By widening the slit they found that the shapes of prominences could be seen directly, at any time, and it was not, therefore, necessary to wait for a total eclipse in order to study them. The emission line spectrum from the chromosphere itself was observed two years later by the American astronomer Charles Young.

In 1892 George Ellery Hale and, practically simultaneously, H. Deslandres, devised an instrument called the "spectroheliograph", which allowed the whole disc of the Sun to be photographed in light of one particular wavelength, such as the "hydrogen-alpha" ($H_\alpha$) line at 656.3 nm, in the red part of the spectrum (*see* diagram 1). (This is one of the wavelengths at which chromospheric hydrogen emits light.) The spectroheliograph uses a series of prisms or diffraction gratings to produce a high dispersion spectrum (one in which the wavelengths are widely spread out). A narrow slit selects a thin strip of the solar disc for observation, while a second slit selects light of one wavelength and allows only this light to fall on a photographic plate. If the image of the Sun is allowed to drift across the field of view while the photographic plate is driven at the appropriate rate behind the slit, an image of the Sun in the chosen wavelength of light will be built up; such an image is called a spectroheliogram. By this means chromospheric phenomena and prominences may be studied at any time over the whole solar disc. Alternatively the monochromatic filter, pioneered by the French solar observer Bernard Lyot in 1933, transmits light over a very narrow band of wavelengths (appreciably less than 0.1 nm wide) centered on some wavelength of interest, such as $H_\alpha$. However, the spectroheliograph and the spectrohelioscope (used for direct viewing rather than photography) are more flexible than monochromatic filters, since the wavelength selected for observation can be altered at will.

### The magnetograph

In 1897 the Dutch physicist P. Zeeman showed that a powerful magnetic field would cause single spectral lines to become double (or broadened). This "Zeeman effect" was observed in the spectra of sunspot regions in 1908 by Hale and colleagues using the 46 m tower telescope at Mount Wilson; their results indicated the presence of magnetic fields thousands of times stronger than that of the Earth, a typical sunspot group having a field of about 3,000 gauss (G) compared to less than 1 G for the terrestrial field. In 1952, H. W. and H. D. Babcock of Mount Wilson devised a much more sensitive magnetograph capable of measuring a weak field of about 1 G on the Sun. Magnetographs based on this model maintain regular monitoring of solar magnetic fields, notably from Mount Wilson and Kitt Peak observatories.

1

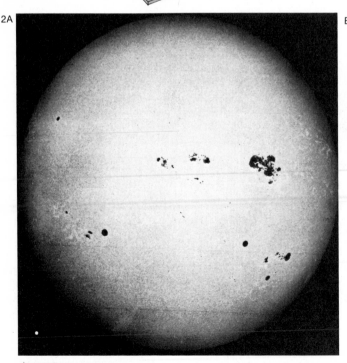

Traveling slit · Image of Sun · Photographic plate · Exit slit · Prism · Mirror

**1. Spectroheliograph**
An image of the Sun is focused onto a narrow slit by means of a telescope. Light passing through the slit passes through a prism, which refracts it, splitting it into component wavelengths. This spectrum is reflected up to a second slit, which selects a single wavelength and allows it to expose a photographic plate. A motor drives the two slits at a steady rate across the image of the Sun, so that an image of the complete disc in the chosen wavelength is gradually built up on the photographic plate.

**2. Spectroheliogram**
A white-light photograph of the Sun (**A**) is shown together with a spectroheliogram (**B**) taken in a band of red light with a wavelength of 656.3 nm known as "hydrogen-alpha". This line was designated as "C" by Fraunhofer and corresponds with the first line in the Balmer series (*see* pages 26–27). This wavelength is emitted principally by the lower part of the chromosphere. Both images were made at Mount Wilson Observatory on 12 August 1917. (The white dot at the lower left is Mercury.)

2A

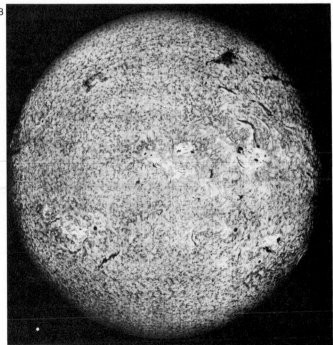

## The coronagraph

Until 1930 it was possible to see the corona only during total eclipses. In that year Lyot constructed a coronagraph—essentially a telescope with a very high quality objective together with a series of internal baffles to cut down background light and a metal disc to block out the photosphere—which allowed the spectrum of the corona to be studied in daylight from high-altitude sites. Subsequent coronagraphs have allowed details of the inner corona to be studied, but it is only since coronagraphs have been placed beyond the atmosphere that it has been possible to maintain continuous coronal observations under perfect conditions.

## The radio Sun

In February 1942 British Army Radar observers picked up a noise signal that J. S. Hey was able to establish came from the Sun, thus demonstrating that the Sun was a radio source that varied in strength. Because of wartime security, these results were not published until 1945. In the meantime, the American amateur Grote Reber constructed a 9.5 m dish antenna and picked up solar radio emission at a frequency of 160 MHz in September 1943, publishing his results the following year. Unsuccessful attempts to detect theoretically predicted solar radio emission had been made by Thomas Edison in 1890 and by Sir Oliver Lodge in 1896. Special observatories such as the Culgoora radioheliograph (*see* page 22) now monitor the Sun's radio activity.

5A
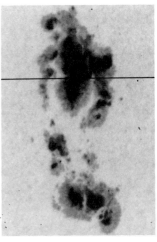
B
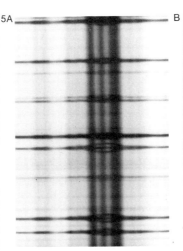

**5. Zeeman effect**
In this photograph of part of the Sun's spectrum (**A**) several lines have been split by the presence of a strong magnetic field. The intensity of the Zeeman splitting indicates a field with a strength of several thousand gauss, compared with the Earth's field of only about 1 gauss. The effect is particularly pronounced in the 525 nm line of iron. The accompanying photograph (**B**) shows the region of the sunspot group at which the spectroscope slit was directed.

**3. Coronagraph**
An image of the Sun is focused by an objective lens onto a field lens, in the center of which is an occulting disc. The occulting disc blocks the image of the photosphere, while the image of the corona is focused onto the photographic plate by a second objective lens. A system of baffles cuts down the stray light reflected inside the coronagraph.

**4. Solar corona**
This image of the corona was produced by the coronagraph carried on board Skylab.

3
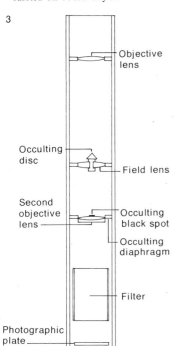

Objective lens

Occulting disc

Field lens

Second objective lens

Occulting black spot

Occulting diaphragm

Filter

Photographic plate

4

# Observatories and Satellites

The Earth's atmosphere seriously hampers observations of the Sun in a number of ways. In particular:

(1) Turbulence causes images to shimmer. The various rays of light making up the image follow different paths through the atmosphere and are distorted in different ways; as a result, much of the potential detail in the image is lost. Sky brightness and variable atmosphere transparency also degrade the quality of observations.

(2) Continuity of observation is interrupted by the day–night cycle and by atmospheric phenomena such as clouds.

(3) Most significant of all, the atmosphere is opaque to most wavelengths of electromagnetic radiation: solar $\gamma$-rays, X-rays, ultraviolet, much of the infrared, and long radio waves cannot be detected from ground level. The Earth's atmosphere is said to have a "window" at those wavelengths that can penetrate as far as ground level. Apart from the "optical" window—visible wavelengths—the main window is in the range of radio waves between about $10^{-2}$ m and 100 m. Particle emissions from the Sun are similarly absorbed and are thus also inaccessible to ground-based observers.

The situation improves with height above sea level and for this reason most solar observatories are built at high-altitude sites such as the tops of mountains, where atmospheric interference will be minimized; among the most important such observatories are Sacramento Peak Observatory, Kitt Peak National Observatory, The Big Bear Solar Observatory (part of the Hale Observatories) and Mount Wilson Observatory. However, the only way of observing the full range of solar radiation is to take instruments above the Earth's atmosphere. As early as 1911 balloons were used to measure solar cosmic rays (heavy atomic particles). The solar spectrum was photographed from an altitude of 22 km in 1935, while magnificent photographs of the photosphere, showing features as small as 150 km across, were achieved in 1957 and 1959 by Princeton University's Project Stratoscope experiment, in which a 25 cm telescope was hoisted to an altitude of 24 km.

Rockets offer the means of attaining much greater altitudes, sufficient to allow solar X-ray and ultraviolet radiation to be observed for a few minutes at the top of the vehicle's trajectory. The solar ultraviolet spectrum was photographed for the first time on 10 October 1946 by a team headed by R. Tousey of the United States Naval Research Laboratory with the aid of a V2 rocket.

Since the launching of the first artificial satellite, Sputnik 1, on 4 October 1957 many satellites have been devoted to exploring the Earth's space environment and the effects of solar radiations, particles and magnetic fields. Other spacecraft have looked directly at phenomena taking place on the Sun's surface and in its atmosphere. Several of the Pioneer series and many of the Explorers supplied data on particles and fields. Specifically devoted to observing the Sun itself were the Orbiting Solar Observatories (OSOs); the first of these, OSO-1, was launched on 7 March 1962 into an orbit at an altitude above the Earth of 560 km. OSO-1 recorded data on some 75 solar flares before its demise on 6 August 1963. Subsequent OSOs extended the range of observations down to 1 nm. The most recent of the series, OSO-8, was launched in 1975. Relationships between the Sun and the Earth have been investigated by satellites such as the Orbiting Geophysical Observatories (OGOs), commencing with OGO-1 in 1964.

In addition to the two major projects, Skylab and the Solar Maximum Mission (*see* pages 24–25), satellites have been launched by NASA on behalf of or in cooperation with other nations, and these have contributed significantly to solar research. Notable among them have been the NASA/West German Helios spacecraft, which passed within 45,000,000 km of the Sun in 1974 and 1976, and the NASA/ESA (European Space Agency) International Sun–Earth Explorers of 1977 and 1978. Various Soviet satellites and the manned Salyut laboratories have also made observations of solar phenomena.

## 1. Spectral windows

The Sun's radiation is selectively absorbed at various levels of the Earth's atmosphere depending on the wavelength of the radiation. This graph shows the altitude at which the intensity of radiation is halved over a range of wavelengths from $\gamma$-rays to radio waves. Visible light and certain other restricted bands of radiation, notably in the infrared and radio portion of the spectrum, can penetrate as far as ground level; these wave bands are said to constitute "spectral windows". Observations at other wavelengths can only be made by means of spacecraft sent outside the Earth's atmosphere.

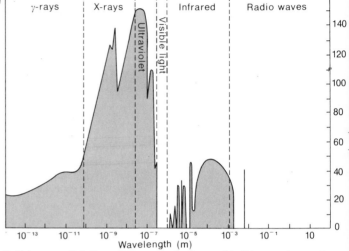

## 2. Culgoora radioheliograph

This aerial photograph shows the radioheliograph site at Culgoora, New South Wales. The instrument was designed specially for solar observations at a frequency of 80 MHz (equivalent to a wavelength of 3.75 m, in the radio portion of the spectrum) and consists of a total of 96 separate parabolic antennas, each with a diameter of 14 m, spaced at equal intervals around a circle 3 km in diameter. The antenna dishes, controlled by a computer, scan the Sun and build up a picture of the radio emissions that come from the solar corona. It is particularly important for the study of flares (*see* pages 66–69). The very large radius of the radioheliograph is necessary to achieve a detailed image at the long wavelength at which it operates: for any telescope, the longer the wavelength, the lower the resolution achieved with a given diameter. An example of an image produced by the Culgoora instrument is shown on page 48.

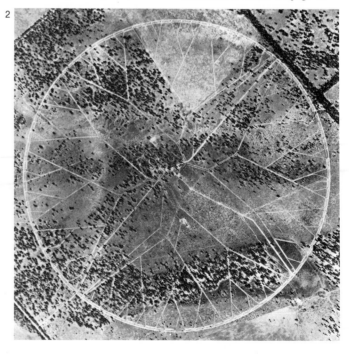

3

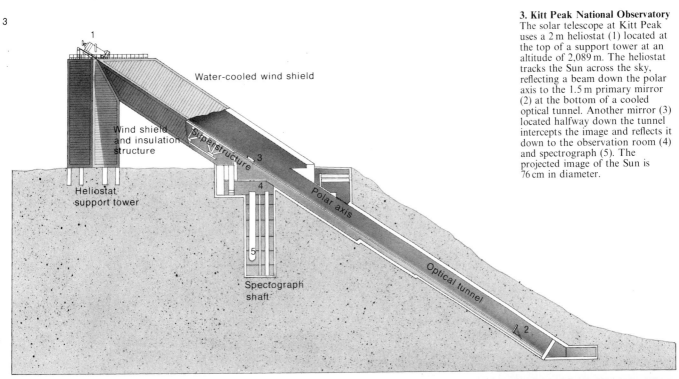

### 3. Kitt Peak National Observatory
The solar telescope at Kitt Peak uses a 2 m heliostat (1) located at the top of a support tower at an altitude of 2,089 m. The heliostat tracks the Sun across the sky, reflecting a beam down the polar axis to the 1.5 m primary mirror (2) at the bottom of a cooled optical tunnel. Another mirror (3) located halfway down the tunnel intercepts the image and reflects it down to the observation room (4) and spectrograph (5). The projected image of the Sun is 76 cm in diameter.

Labels on figure: 1 · Water-cooled wind shield · Wind shield and insulation structure · Superstructure · 3 · Heliostat support tower · 4 · Polar axis · 5 · Spectrograph shaft · Optical tunnel · 2

### 4. Project Stratoscope
This photograph of a sunspot was taken on 17 August 1959 by a camera carried by Princeton University's Project Stratoscope. Photospheric granulation and the details of the penumbra are clearly visible.

### 5. Ultraviolet spectra
The first measurements of the extreme ultraviolet portion of the Sun's spectrum were made by a V2 rocket launched on 10 October 1946. At altitudes up to 34 km wavelengths below about $3 \times 10^{-7}$ m are absorbed by the ozone layer. At 55 km the rocket penetrates the most concentrated region of ozone and shorter wavelengths start to be detected.

### 6. OSO 8
The last of the Orbiting Solar Observatories was launched on 21 June 1975. It carried ultraviolet telescopes to investigate energy transfer between different layers of the Sun, and a special arm to detect rapid X-ray increases.

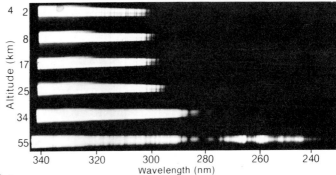

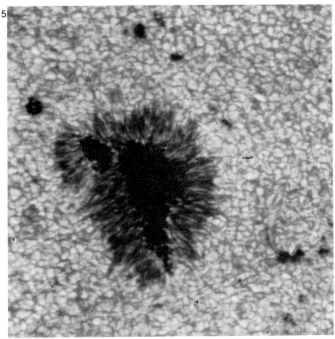

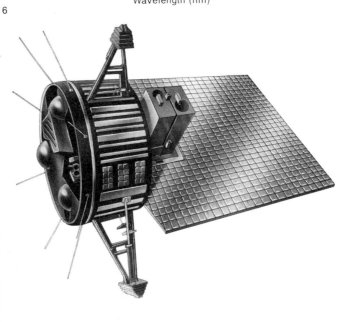

# Skylab and Solar Maximum Mission

The largest structure so far to have been placed in space, the American manned orbiting laboratory Skylab, was set into orbit at an altitude of 435 km on 14 May 1973. Crews were ferried to and from the laboratory by means of Apollo-type spacecraft. With the Apollo Command Service Module (CSM) attached, Skylab had an overall length of 36 m and a mass of 90,600 kg. The major component, the orbital workshop, provided a total of 292 m³ of living space, and three crews, each of three astronauts, spent a total of 513 man-days in space, carrying out a wide variety of experiments and observations; solar astronomy occupied about 30 percent of the total time allocation. The astronauts were able to coordinate their experiments closely with the work of ground-based observatories, with whom they maintained contact constantly.

Eight different instruments were mounted on the Apollo Telescope Mount (ATM), a structure which allowed the instruments to be held rigidly and pointed with high accuracy towards the Sun. The basic ATM structure measured 4.4 m tall, weighed 11,090 kg, and was powered by solar panels arranged like windmill sails 31 m in diameter. The instruments comprised a white light coronagraph (which allowed the corona to be studied to a distance of about 6 $R_S$), three X-ray instruments, which produced X-ray photographs and spectra, three ultraviolet instruments, which supplied images and spectral data, and two $H_\alpha$ telescopes, permitting direct observation and photography of hydrogen light phenomena and flares. In addition, there was a hand-guided X-ray/ultraviolet solar photography experiment operated through an airlock in the spacecraft's laboratory.

Skylab itself reentered the atmosphere and broke up on 11 July 1979. Ironically, this occurred as a result of the effects of increasing solar activity on the uppermost regions of the Earth's atmosphere. The contribution of Skylab to knowledge of the Sun was immense, including the discovery of important new phenomena such as coronal holes (see pages 74–75).

## Solar Maximum Mission (SMM)

Designed primarily to make a concentrated attack on problems associated with the nature, trigger-mechanism and effects of solar flares, this satellite was launched on 14 February 1980 into a circular orbit at an altitude of 574 km. It contained a battery of carefully matched instruments operating in white light, ultraviolet, X-ray and $\gamma$-ray ranges. Other objectives included the study of the evolution of the corona around a maximum period of solar activity, and the accurate measurement over a prolonged period of the total solar output of radiation. At the time of writing (one year after launch) the mission has unfortunately lost much of its gyroscopic pointing capability; however, the vehicle is of a type suitable for repair or recovery at a future date by means of the Shuttle.

The spacecraft itself has a mass of 2,315 kg and measures 4 m long by 1.2 m wide. It consists of an instrument module, housing all seven solar instruments (see table), together with the Fine Pointing Sun Sensor used to lock the instruments onto the Sun, as well as the supporting spacecraft, which itself comprises three modules of essential subsystems: attitude control, power and communications, and data handling. Two arrays of solar panels supply about 3,000 W of power for the spacecraft's systems, while battery power is also available for the periods when the spacecraft is in the Earth's shadow, or in eclipse.

The unique feature of this mission is the degree of integration that exists between the various instruments, and the degree of flexibility that is built into the mission operations. The investigators associated with each experiment are housed together in the Experimenters' Operations Facility at NASA's Goddard Space Flight Center and meet daily to decide, on the basis of current observations, which active area to concentrate their efforts on during each 24 hr period. In this way the onboard equipment can be used with the greatest efficiency to study such transient phenomena as flares.

**1. Solar Maximum Mission**
The spacecraft carried seven scientific instruments: (1) X-ray Polychromator, designed to investigate the activity that produces solar plasma with temperatures between 1.5 and 50 million K; (2) Solar Irradiance Monitor, to measure variations in the solar constant (see pages 80–81); (3) Coronagraph/ Polarimeter, producing images of coronal evolution and coronal transient activity; (4) Hard X-Ray Burst Spectrometer, to investigate the role of energetic electrons in solar flares; (5) Ultraviolet Spectrometer and Polarimeter, to study coronal active regions and flares by observing spectral features in the UV range; (6) $\gamma$-Ray Spectrometer, to examine ways in which high-energy particles are produced in solar flares; (7) Hard X-Ray Imaging Spectrometer, to provide information about the position, extension and spectrum of hard X-ray bursts in flares.

**2. Skylab**
Skylab had a length of 25 m and was 6.7 m wide at the main workshop section. While in orbit its instruments were powered by solar cells. (One of the large solar panels shown in this illustration was torn off during launch, and photographs show it with only one in position.)

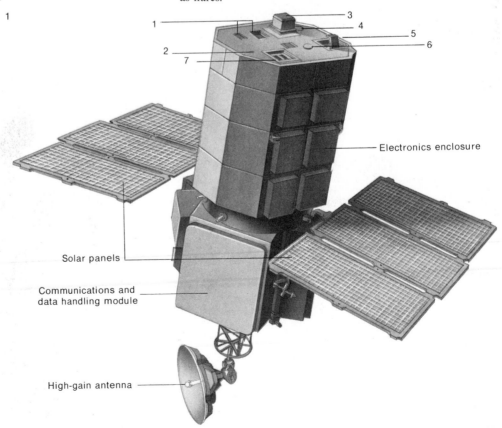

Electronics enclosure

Solar panels

Communications and data handling module

High-gain antenna

1. Modified Apollo Command Module and Service Module
2. Service Propulsion Engine
3. Radiators
4. Attitude Control Jets
5. Crew Station
6. Apollo Telescope Mount
7. Solar Cells
8. Sun Shield
9. Telescope Apertures
10. Oxygen Tank
11. Nitrogen Tank
12. Maneuvering Unit

13. Gravity Substitute Workbench
14. Food Provisions
15. Solar Cells
16. Sleep Restraints
17. Water Containers
18. Aerial
19. Multiple Docking Adapter
20. Alternative Docking Port
21. Atmosphere Interchange Duct
22. Descent Battery Packs

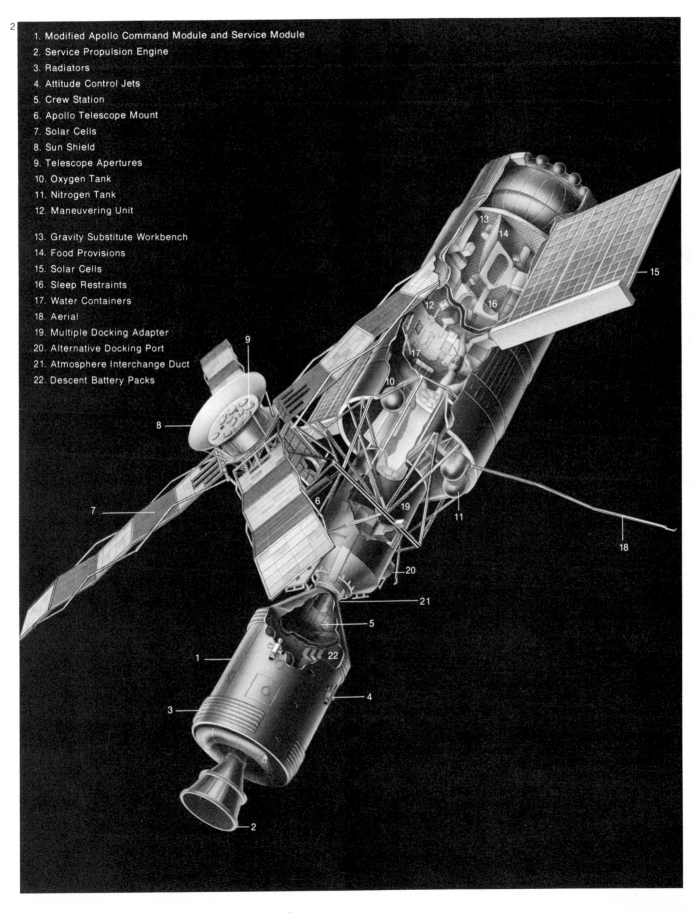

# The Solar Spectrum

Most of what is known about the Sun is derived from analysis of its spectrum. The Sun's visible spectrum, like that of most other stars, consists of a continuum (continuous spectrum) with dark absorption lines. An oversimplified view is that the continuum is emitted from the photosphere, and that the absorption lines are produced as this radiation passes through the more rarefied solar atmosphere above the base of the photosphere. Indeed, it used to be thought that there was one particular layer, called the "reversing layer" where these lines were formed. In fact emission and absorption take place throughout the solar atmosphere (photosphere and chromosphere), but as far as visible light is concerned the effects of absorption become more important with increasing height above the base of the photosphere.

## The electromagnetic spectrum
Visible light spans a range of wavelengths from about 400 to about 700 nm, but the visible or optical range of wavelengths represents only a tiny band in a much wider range of wavelengths making up the entire spectrum of electromagnetic radiation. Electromagnetic waves of all kinds travel through a vacuum at a constant speed, the velocity of light, equal to $299,792.5 \text{ km s}^{-1}$. They range in wavelength from a microscopic fraction of 1 nm to many kilometers; the complete range of wavelengths has been divided arbitrarily (from shortest to longest wavelength) into $\gamma$-ray, X-ray, ultraviolet, visible, infrared, microwave and radio. Other subdivisions encountered in solar astronomy include extreme ultraviolet (EUV), which spans from 10 nm to about 120 nm, soft X-ray (0.1 to 10 nm), and hard X-ray (less than 0.1 nm), while at wavelengths longer than visible (10 $\mu$m to 1 mm) the term "far infrared" is used, and the term "radio" is often used for any wavelength longer than 1 mm.

## Black body radiation
A "black body" is the name given to an idealized object that is a theoretically perfect emitter of radiation. Such a body absorbs all radiation falling on its surface and reflects nothing; it will also emit, in a characteristic way, all the energy that is supplied to it; in other words, a good absorber is also a good emitter. A black body will emit all kinds of radiation, but the amount of energy emitted at different wavelengths depends upon its temperature.

A graph of the amount of energy plotted against wavelength follows a distinctive curve called a "black body distribution curve" (*see* diagram 1); for any given temperature there is one particular wavelength of peak emission, and the intensity of emitted radiation drops off sharply at shorter wavelengths and more gradually at longer ones. According to the Wien displacement law, the wavelength of peak emission is inversely proportional to temperature: the higher the temperature the shorter the wavelength of peak emission. The color of a hot body, which depends on the wavelength of peak emission, is therefore determined by its temperature: in everyday terms, when a lump of metal is heated it first glows dull red, then as the temperature increases it becomes progressively orange, yellow and white hot. A body at room temperature (roughly 300 K) emits most strongly in the infrared at a wavelength of about 10 $\mu$m while a 6,000 K black body would radiate mainly in the visible, peaking at about 550 nm in the yellow-green part of the spectrum. A body at 30,000 K would have its peak of emission at around 100 nm—well into the ultraviolet.

Although stars are not ideal black bodies, their continuum emissions can broadly be described by black body curves. The "effective temperature" of a star is the temperature that a black body of the same radius as that star would need to have in order to emit the same *total* quantity of energy as the star. The effective temperature of the Sun, for example, is 5,780 K. As a result, the Sun emits most strongly in the middle of the visible range. Observed from a distance, the Sun would appear as a yellow star.

## The absorption lines
The solar spectrum reaching the Earth includes thousands of dark lines originating in the solar atmosphere. It is further complicated by additional absorption lines produced in the Earth's atmosphere called "telluric lines".

The way in which absorption lines are produced depends upon the physics of the atom. The simplest atom, hydrogen, may be visualized as consisting of a central proton (a heavy particle of

**1. Black-body radiation**
The distribution of energy flux against wavelength of a radiating black body follows a distinctive curve: for any given temperature the curve is unique. As the temperature of the body rises the total amount of energy emitted increases rapidly, and the wavelength of peak emission shifts towards the blue (as shown by the dotted line).

**2. Bohr atom**
A simplified model of the hydrogen atom proposed by Niels

Bohr helps to account for the production of spectral lines. The atom's electron is represented as orbiting the nucleus; only certain orbits are possible, each with its own energy level. When an electron makes a transition from a higher to a lower level, a photon (whose energy corresponds to the difference between the two levels) is emitted, producing an emission feature; conversely, absorption features are produced by the absorption of photons.

**3. The Solar Spectrum**

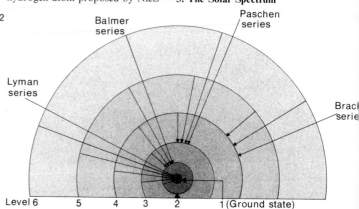

Wavelength (nm)

violet
blue
blue-green
green
red
orange
yellow
yellow-green

6,000 K
5,000 K
4,000 K
3,000 K

Balmer series
Paschen series
Lyman series
Brackett series

Level 6   5   4   3   2   1 (Ground state)

positive electrical charge) around which there circulates a single electron (a lighter particle of negative charge). According to quantum theory, this electron may exist only in one of a number of orbits, each of which corresponds to a certain energy level of the atom; the atom has least energy when the electron is in the lowest orbit—the "ground state". In some respects light behaves like a wave but in others it behaves like a beam of particles; a "particle" of light, or "photon", has a certain amount of energy corresponding to the wavelength of the light. In fact the energy is inversely proportional to wavelength, so that the shorter the wavelength, the higher the energy of the photon.

If the energy of an incoming photon corresponds to the energy difference between one electron orbit and another, then the photon may be absorbed by an atom. When a photon is absorbed in this way, the electron jumps (makes an "upward transition") to a higher orbit. On the other hand, an electron in a high orbit can fall down to a lower one, and the difference in energy is released in the form of a photon with a wavelength that again corresponds to the energy difference between the two orbits. In essence, then, absorption of light occurs when electrons make upward transitions, and emission occurs when they make downward transitions.

Many different transitions can take place in a hydrogen atom (*see* diagram 2). For example, an electron falling from level 3 to level 2 releases red light at a wavelength of 656.3 nm, while an electron falling from level 4 to level 2 emits at a wavelength of 486.1 nm. The greater energy difference corresponds to a shorter wavelength. The complete series of lines corresponding to all possible transitions to level 2 is known as the Balmer series, the lines being labelled $H_\alpha$ (at 656.3 nm), $H_\beta$ (486.1 nm) and so on. Other series occur in different regions of the spectrum. The Lyman series, for example, involving transitions to the ground state, is found in the ultraviolet, while the Paschen and Brackett series occur in the infrared.

Each chemical element has its own characteristic series of lines which allows its "fingerprint" to be recognized in the solar spectrum. The detailed analysis of the spectrum, however, is complicated by many factors. Heavier atoms have more electrons and can exist in various states of "ionization" (an atom is said to be ionized when it has lost one or more electrons). The state of ionization is indicated by means of Roman numerals; for example, ordinary neutral hydrogen is denoted by HI while ionized hydrogen, which has lost its one electron, is represented by HII, and iron which had lost, say, 13 electrons would be symbolized by FeXIV. The level of ionization affects the pattern of absorption features so that it is possible to determine from an examination of the spectrum not only which elements are present, but also their state of ionization. This information in turn can be used to measure the temperature of the gas because extremely high levels of ionization are known to be produced at very high temperatures. Molecules, moreover, have complex spectra.

## Doppler shift

In practice a spectral line is not a perfectly black line at one precise wavelength. If the intensity of its absorption of the continuum is mapped against wavelength, it is usually found that the line has a shape, or "profile", dipping down slowly at first and then sharply as the central wavelength of the line is approached; with increasing wavelength the amount of absorption drops off sharply at first and then more slowly. The line appears to have a central core with "wings" on either side (*see* diagram 2). One of the factors responsible for spreading out the line over a range of wavelengths is the "Doppler effect": the movement of a source relative to an observer causes a shift in the observed wavelength. In a gas, the motions of the absorbing atoms, due essentially to the temperature of the gas, cause small changes in the wavelengths at which absorption is observed. An approaching atom will be seen to absorb (or emit) at a shorter wavelength than a stationary atom, while a receding atom will absorb or emit at a longer wavelength. The random motion of atoms in a hot gas ensures a spread of wavelengths on either side of the mean value.

Other factors also play a part. Pressure broadening, whereby collisions between atoms and electrons change the perceived wavelength, the presence of electric and magnetic fields (which can lead to the splitting of the line into two or more components) and other factors all contribute to the line profile (*see* pages 34–35).

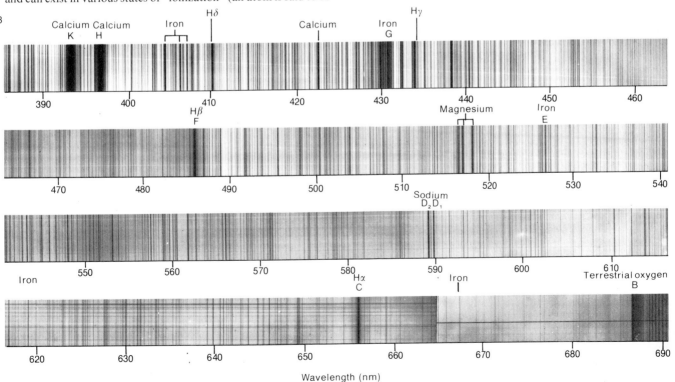

Wavelength (nm)

# Solar Radiation

Analysis of the solar spectrum can reveal a wealth of information about the Sun and its physical properties. By investigating the spectrum, for example, it is possible to deduce facts about the conditions at different heights above the base of the photosphere: at any given wavelength it is possible to "see" down into the solar atmosphere only as far as the level at which it becomes opaque to that particular radiation. Bearing in mind that a good absorber is also a good emitter, this implies that the greatest part of the radiation received at a given wavelength originates from the level at which the solar atmosphere becomes opaque to that wavelength (the "emission level"). From below that level, the radiation is absorbed too effectively to reach Earth, while above that level the solar atmosphere is transparent, absorbing and emitting very little.

At the central wavelength of a very dark absorption line the solar atmosphere is highly opaque, and so any radiation received must originate from a relatively high level; towards the "wings" the absorption is less and the radiation received originates at a lower level. The continuum radiation comes from the low photosphere—at least in the visible part of the spectrum. Ultraviolet radiation is received primarily from the chromosphere, the photosphere being opaque to most of it. At the shortest radio wavelengths (in the millimeter and centimeter ranges) radiation is received from just above the photosphere, but at longer wavelengths the solar atmosphere is opaque at higher levels; at meter wavelengths the radiation comes from the low corona.

### Radio emissions from the Sun
Although the black body curve for a body with a temperature of around 6,000 K, continued into the radio region, would suggest that the Sun ought to be a very weak radio emitter, in fact observations show that the radio output of the Sun is a highly variable quantity that can flare up at certain wavelengths by factors as large as 10,000.

Solar radio emission consists of three main components:
(1) The "quiet Sun" emission is the total background output of the Sun, excluding discrete (localized) sources, and takes the form of "thermal" (black body) radiation emitted from the randomly moving particles in a hot gas. The apparent "brightness temperature" of this radiation ranges from about 6,000 K at millimeter wavelengths to over $10^6$ K at meter wavelengths, the different temperatures relating to different levels in the atmosphere. (The brightness temperature associated with radiation of a particular wavelength is the temperature a black body of the same size as the Sun would need to have in order to emit the measured quantity of radiation at that wavelength.)
(2) The slowly varying component (S-component) is again thermal radiation, but is emitted from localized regions of the solar atmosphere, the total quantity of radiation emitted by the whole Sun seldom much exceeding the "quiet Sun" level, but being locally intense and depending on the level of solar activity. It is most prominent in the "decimeter" wavelength range between about 10 and 50 cm (frequencies of around 3 GHz to 600 MHz).
(3) Radio bursts can occur over the entire radio spectrum over timescales ranging from less than 1 sec up to several hours. Their power outputs can exceed the quiet Sun level by factors of 1,000 to 10,000, the radiation being mainly of a nonthermal nature: it is emitted by electrons which, instead of moving at random as in a hot gas, are directed in their motions, under the influence of, for example, magnetic fields. One important type of nonthermal process is "synchrotron radiation": electrons moving at a large fraction of the speed of light in strong magnetic fields are forced to spiral around the magnetic lines of force and, as a result, emit radiation in a narrow cone along the direction in which they are moving. Solar radio radiation is polarized (see Glossary), usually due to the role of the magnetic field.

Despite the spectacular nature of these events, the overall contribution of the radio bursts to the total radio output over a long period is quite small.

### Infrared radiation
At wavelengths from about 750 nm to just under 1 mm the observed radiation originates in the photosphere and lower chromosphere

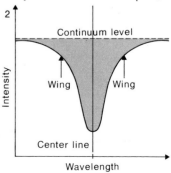

**1. Spectrum of solar radiation**
As can be seen from this graph, the peak intensity of solar emission is in the narrow band of visible wavelengths. At wavelengths between about 100 nm and 1 cm, radiation from the Sun corresponds quite closely to the radiation that would be emitted by a "black body" with a temperature of 6,000 K. (The curve for a black body at this temperature is shown as a dotted line in the graph.) At shorter wavelengths, extending from the ultraviolet to the shortest $\gamma$-rays, and at very long wavelengths, solar radiation deviates significantly from this pattern: the black body temperature to which these extremes of emission correspond is closer to $10^6$ K. Unlike the central portion of the spectrum, the intensity of radiation in the short and long wavelengths varies considerably, depending on the level of solar activity. At radio wavelengths, for example, there is a slowly varying component which changes with the solar cycle (see pages 40–41), while different types of flare (see pages 66–69) produce distinctive emissions; the strong emission lines that appear in the $\gamma$-ray region are shown in the graph.

**2. Absorption line profile**
If a graph is plotted of intensity against wavelength, it is found that the cut-off of an absorption feature is not abrupt but slightly spread out over a range of wavelengths. In the visible spectrum the center of the line is, generally, formed higher in the atmosphere than the "wings" of the line, which are formed at deeper regions where the temperature (and consequently the level of emission) is relatively high. Thus, by making monochromatic images of the Sun at wavelengths slightly off the centers of absorption lines it is possible to observe a range of depths into the solar atmosphere.

and follows quite closely the black body curve for a body in the temperature range 6,000 K to 4,000 K (the lower temperatures correspond to radiation coming from the region of the temperature minimum some 500 km above the base of the photosphere).

## Ultraviolet radiation
In the near ultraviolet (around 300 nm) the radiation originates, like visible light, in the photosphere, but at progressively shorter wavelengths the source of the ultraviolet continuum shifts to the chromosphere. At wavelengths shorter than about 140 nm the solar material is sufficiently opaque to radiation from lower levels for it to become possible to see emission lines from the chromosphere itself. Thus, below that wavelength, the spectrum changes from a bright continuum with dark absorption lines to a faint continuum with bright emission lines. Studies of the Sun in the wavelengths of ultraviolet emission lines allow higher and higher layers of the chromosphere and transition region to be investigated. One particularly bright line is the hydrogen line Lyman-alpha ($L_\alpha$, the first line of the Lyman series) at 121.6 nm which emits more energy than the entire solar spectrum between 0 and 120 nm; this radiation exerts important effects on the Earth's atmosphere (*see* pages 78–79) and on other Solar System phenomena.

## Extreme ultraviolet (EUV) and soft X-ray radiation
Radiation in the 1 nm to 120 nm range also has important terrestrial effects. It originates in the chromosphere, transition region and corona, and at wavelengths of 1 to 2 nm can show variations in intensity over the solar activity cycle of factors in excess of 100. The soft X-rays are emitted by hotter, denser concentrations of gas in the corona (with a temperature in excess of $10^6$ K) and a study of these radiations shows the detailed structure of the corona across and beyond the solar disc.

Hard X-rays (wavelengths less than 0.1 nm) originate primarily in solar flares, the intensity of such emission fluctuating widely by as large a factor as 10,000 depending on the level of solar activity. These outbursts have significant effects on the Earth's atmosphere.

## The solar constant
The total amount of energy per second over all wavelengths that would be received at the top of the Earth's atmosphere when the Earth is located at its mean distance from the Sun is called the "solar constant". Although this quantity is of the utmost importance to life on Earth (*see* pages 80–81), surprisingly its value is not yet known to an accuracy of better than 0.5 percent, and it is unclear by how much—if at all—the value of this "constant" varies with time. The difficulty arises mainly because, until comparatively recently, all measurements had to be made from the Earth's surface and allowance had to be made (often incorrectly) for the fact that the atmosphere obscures many wavelengths and exerts a variable absorption on the others. There is also a problem of calibration, since it is hard to find a reference standard for a source as bright as the Sun. One of the principal experiments on the Solar Maximum Mission is devoted to improving knowledge of this quantity and assessing variability (*see* page 24).

The presently accepted value of the solar constant is about 1,370 W m$^{-2}$; in other words, a surface of area 1 m$^2$ held perpendicular to the Sun's rays should receive 1,370 W of power at the mean distance from the Sun to the Earth. There is an uncertainty of nearly $\pm$ 10 W on this figure—the mean of values measured for the first 153 days of the SMM was 1,368 W m$^{-2}$ (*see* pages 80–81).

Overwhelmingly the greatest part of solar radiation is emitted in the visible and near infrared part of the spectrum, and so the major contribution to the solar constant is in those wavelengths. In fact, about 99 percent of solar radiation is emitted in the wavelength range 300 nm to 6,000 nm; just under 1 percent is contributed by ultraviolet radiation in the range 120 nm to 300 nm, and all other wavelengths contribute only a tiny fraction of 1 percent. The solar output in those wavelengths that contributes significantly to the solar constant does seem to be constant to within 1 percent. The radio and X-ray wavelengths, which show huge fluctuations, contribute insignificantly to the total. A variation of $\pm$ 3.5 percent from the average value arises annually, due to the elliptical form of the Earth's orbit.

**3. Radio map of the Sun**
This contour map shows the Sun at a wavelength of 3 cm. Active regions appear as the sources of enhanced radio emission.

**4. Synchrotron radiation**
When an electron moving at a "relativistic" velocity (a large fraction of the speed of light) spirals around magnetic field lines, it emits a narrow beam of strongly polarized radiation. The wavelength is dependent on the strength of the field and the velocity of the electron. This process is an important source of radio bursts emitted by the Sun as well as by many other astronomical radio sources.

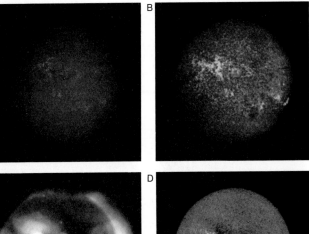

Radiation

Electron

Magnetic field lines

3

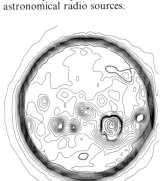

**5. Contrasting solar images**
Spectroheliograms are used to show the Sun at various wavelengths, such as Hα at 656.28 nm (**A**) or the K-line of doubly ionized calcium at 393.37 nm (**B**). (**C**) is an X-ray image, while (**D**) is a magnetogram.

5A

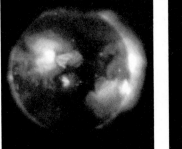

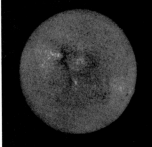

# The Photosphere

The photosphere is the layer from which practically all the Sun's visible light is emitted, and as such represents what is normally called the "surface" of the Sun. Although the edge or "limb" of the solar disc appears to be quite sharp in a photograph or in the projected solar image as seen with a telescope (apart from the "boiling" effect due to turbulence in the atmosphere), it is fairly readily apparent that the brightness of the solar disc diminishes near its edge. This phenomenon of "limb darkening" is consistent with the Sun's being a gaseous body having a temperature that decreases with distance from its center.

Since the Sun is gaseous, its photosphere is to some extent transparent: it can be observed down to a depth of a few hundred kilometers before it becomes completely opaque. When an observer looks directly at the center of the solar disc, he is in fact looking through photospheric layers of increasing density and temperature towards the base of the photosphere, where the solar material becomes opaque. Most of the light he sees at the center comes from the hotter, and therefore more luminous, lower layers of the photosphere. However, looking towards the limb, an observer's line of sight is almost at a tangent to the edge of the solar disc, so that he is looking into the cooler, upper layers of the photosphere. Thus although it is again possible to penetrate a certain distance into the photosphere before the solar material becomes opaque, the region at which this occurs is higher above the base of the photosphere than at the center of the disc. The light from the region of the limb, therefore, comes from higher layers which, being at a lower temperature, emit less strongly than the deeper layers at the base of the photosphere: consequently the limb appears less bright than the center.

The degree of limb darkening is not the same for all wavelengths. It is most apparent in the blue and violet regions of the spectrum, because the intensity of blue light diminishes with decreasing temperature more rapidly than does the intensity of red light (*see* pages 26–27). At radio wavelengths and at shorter ultraviolet and X-ray wavelengths the opposite effect, "limb brightening", is observed. These types of radiation emerge primarily from layers in the solar atmosphere well above the photosphere, in the regions where temperature increases with height instead of decreasing as it does in the photosphere. As a result the radiation from the limb regions (which again become opaque at higher levels than the central regions) comes from *brighter* high-temperature layers. The limb therefore appears brighter than the center. In the case of X-ray sources in the corona, absorption is not important but there are more X-ray emitting sources to be seen along the line of sight when looking beyond the edge of the Sun than when looking towards the center; again, the limb regions appear brighter.

The pressure and density of the gas in the lower layers of the photosphere are, respectively, only about 10 percent and 0.01 percent of the values pertaining to the Earth's atmosphere at sea level. Terrestrial air is highly transparent to visible light; why should the photospheric material become opaque at such low densities and pressures? The *opacity* of a gas is a measure of its ability to absorb light, and the value of this quantity depends upon the state in which the gas exists. For example, a gas made up of neutral hydrogen atoms in their lowest energy states interacts very weakly with photons, and so is highly transparent; on the other hand, a hotter hydrogen gas is more ionized and excited, and interacts strongly with optical photons, absorbing them very effectively. The deep interior of the Sun is completely ionized and highly opaque.

The photospheric material is opaque mainly as a result of the presence of negative hydrogen ions ($H^-$) (hydrogen atoms that have captured an extra electron). A photon emerging from the interior which encounters an $H^-$ will be absorbed, and the extra electron will be ejected in the process. When at a later moment the resulting neutral hydrogen atom recaptures an extra electron a photon is

**1. Limb darkening**
The apparent fall in brightness away from the center of the Sun's disc is evident in this white light photograph taken on 9 April 1970. The solar atmosphere is partially transparent, and at the center of the disc it is possible to see through the upper layers down to a region where the temperature is higher and emission correspondingly brighter; at the limb, however, high temperature layers do not fall along the oblique line of sight and the limb therefore appears darker.

**2. Granulation**
The textured appearance of the photosphere, described as "solar granulation", results from the turbulent motion of hot gas rising and falling as heat is transported from the interior by means of convection. Each granule is about 1,000 km across and represents the top of a rising column of hot gas. Downward motion occurs in the dark lane between cells. The mean lifetime of a granular cell is approximately 10 min. This photograph was taken on 19 August 1975.

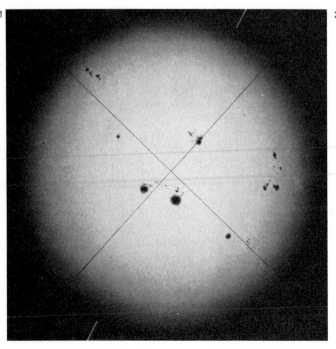

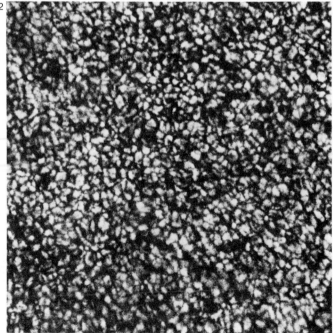

emitted, possibly with a different wavelength from the originally "captured" photon. In this way radiant energy from the interior is processed by the photosphere and emerges as visible light. The concentration of H⁻ drops off very rapidly with decreasing temperature, so that with increasing height above the base of the photosphere the opacity drops very sharply and emitted photons have increasing chances of escaping into space without being captured again. The drop in opacity is so rapid that the major part of the emitted light comes from a layer only about 100 km thick.

### Granulation and supergranulation

The whole photosphere is a seething mass of bright, shifting granules (*see* pages 14–15) whose average lifetime is about 8 min. The individual granules are typically about 1,000 km in diameter, corresponding to angular sizes, as seen from the Earth, of between 1 and 2 seconds of arc. Theoretically such features should be discernible in a telescope of about 10 cm aperture, but the turbulence of the atmosphere normally precludes this. Each granule represents a region in the center of which hot gas is rising from the interior (with a vertical velocity of about $0.5 \text{ km s}^{-1}$), then spreading horizontally outward at the top of the cell at speeds of the order of $0.25 \text{ km s}^{-1}$. Cooler gas descends at the edge of each cell in the intergranular lanes.

On a much larger scale there is a network of "supergranular" cells with typical cell diameters of about 30,000 km; each supergranular cell contains hundreds of individual granules. Being larger

patterns of circulation that extend deeper into the convective zone they survive longer than ordinary granules, with typical lifetimes of between 12 and 24 hr. The solar magnetic field is enhanced around the boundaries of supergranular cells, giving rise to the "chromospheric network" in the chromosphere (*see* page 32).

The pattern of supergranular cells is readily revealed in what are known as "velocity-cancelled spectroheliograms". By the Doppler effect, light from a source approaching the observer is blue-shifted (shortened in wavelength) while light from a receding source is red-shifted (lengthened in wavelength). Since supergranular cells exhibit a horizontal motion of gas out from their centers, it follows that some of the gas will be approaching the observer and matter in other parts of each cell will be receding. If one spectroheliogram is made at a wavelength a little to the red side of the center of a spectral line and another is made a little to the blue side, the red-shifted and blue-shifted regions of gas will be revealed. The pattern of motions is enhanced if the two spectroheliograms are combined photographically in an appropriate way, revealing the supergranular cells. Corrections are made to allow for solar rotation.

Telescopically, the most obvious signs of activity on the photosphere are the sunspots (*see* pages 36–43). Associated with sunspots, however, are "faculae", brighter than average patches of light in the upper regions of the photosphere which usually appear in the vicinity of a sunspot group before the sunspots themselves emerge, and which usually persist for several weeks after the disappearance of the spots.

**3. Doppler effect**
The movement of a source of light relative to the observer produces an apparent shift in the wavelength (or color) of the light seen by the observer. If the source is moving towards observer A then it will appear that the distance between successive crests is compressed, producing a shift towards blue; if, instead, the source is moving away to observer B, the distance between crests will appear longer and the light will look redder than if the source had been stationary.

**4. Supergranulation**
Large-scale convective motion produces "supergranular" cells similar to granulation but about 180 times larger and with longer lifetimes. Supergranulation is shown up in this velocity-cancelled spectroheliogram, in which receding areas appear as dark regions and approaching areas as light regions.

**5. Faculae**
Bright patches in the photosphere, or faculae, can be seen near the limb of the Sun.

Observer A

3 Direction of travel

Shorter waves (Blue shift)

Longer waves (Red shift)

Observer B

4

5

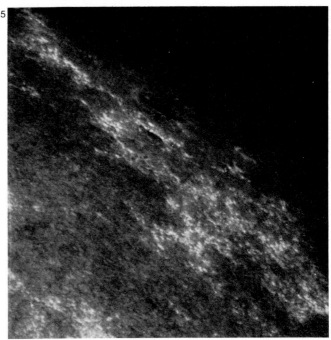

# The Chromosphere I

The chromosphere is the tenuous layer of gas, a few thousand kilometers thick, which lies above the photosphere and which derives its name (literally, "color sphere") from the characteristic reddish-pink color it displays when it becomes visible for a few seconds at the onset and at the end of the total phase of a solar eclipse. Far too faint to be seen against the brilliance of the photosphere, the chromosphere may be observed visually only during a total eclipse, but it can be studied at any time with the aid of a spectroscope, the slit of which is placed at a tangent to the solar disc, when observations are made in one of the wavelengths at which chromospheric matter emits light. The chromosphere may be examined across the entire solar disc by means of a spectroheliograph or a monochromatic filter (*see* pages 20–21).

### Spicules

The chromosphere is not a homogeneous atmospheric layer, but has considerable structure and shows rapidly changing aspects both locally and on the large scale. In monochromatic light, the edge of the Sun is seen to be made up of large numbers of flame-like protrusions, known as "spicules", which rise and fall, looking like blades of grass blowing in the wind; their appearance has been described as a "burning prairie". Spicules are near-vertical cylinders of chromospheric gas which follow the directions of the local magnetic fields. Their temperatures typically range from 10,000 to 20,000 K; they are about 1,000 km broad and rise to heights of about 10,000 km. These jets of gas rise from the lower chromosphere with velocities of 20 to 30 km s$^{-1}$, most of their material eventually falling back and, possibly, contributing to the heating of the chromosphere in the process. Their average lifetimes are between 5 and 10 min.

According to one theory the energy stored in localized magnetic fields is responsible for throwing up spicular material; there is no doubt that local fields influence the paths along which this material moves. Spicules are seen as bright emission features at the limb, but it is possible to see them as dark absorbing features against the disc. At any one time there may be about 500,000 of them on the complete solar surface.

### Chromospheric network

The chromosphere exhibits a large-scale cellular structure known as the "chromospheric network". The cells making up the network coincide in position with the supergranules of the photosphere, their boundaries being determined by the enhanced magnetic fields that occur at the perimeters of supergranular cells. In the photosphere the density of matter is sufficiently great for the flow of hot plasma to sweep the solar magnetic fields along with it. The outward horizontal flow of material from the centers to the edges of supergranular cells thus sweeps the magnetic fields to the boundaries of the cells where they are concentrated and amplified; vertical magnetic fields of 1,000 to 2,000 G are found at the corners. The lower density plasma of the chromosphere is governed in its behavior by the magnetic fields acting upon it; hence the chromospheric network is determined by the underlying magnetic fields. Spicules rise up in bunches in the concentrated fields at the corners of the network.

The network takes the form of a pattern of bright emission features seen in monochromatic light. The pattern is not normally seen at altitudes greater than about 10,000 km above the photosphere (*see* page 51); beyond this level, the clumps of spicules at the cell boundaries spread out, blurring the pattern.

Skylab ultraviolet observations show that the chromosphere takes on a different appearance in polar regions. The chromospheric network cuts off at the edges of these regions, and darker polar caps are revealed coinciding with long-lasting "coronal holes" out of which solar wind particles are streaming (*see* pages 76–77). The height of the transition region jumps up by as much as

1

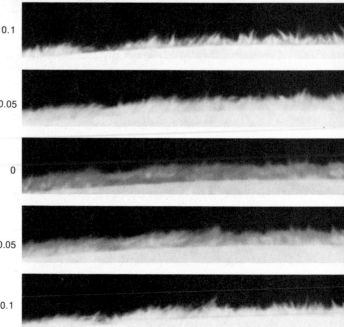

2

+0.1

+0.05

0

−0.05

−0.1

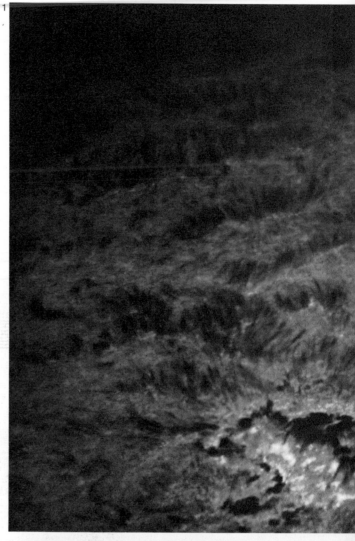

## 1. Spicules

When seen at the limb, spicules are visible as emission features, but against the relatively bright solar disc they show up as dark blade-like absorption features. Bunches of spicules at the edges of supergranular cells outline the chromospheric network. Some distance from the limb, the bunches of spicules make up patterns known as "rosettes"; the bases of these rosettes are marked by small bright mottles. Closer to the limb the rosettes are foreshortened so that the bunches of spicules look like "bushes". The material comprising spicules rises up from the lower chromosphere at speeds of some 20 to 30 km s$^{-1}$, reaching heights of about 10,000 km before falling back. This filtergram was taken on 13 February 1971 at a wavelength 0.08 nm longer than the central wavelength of Hα.

## 2. Spicule sequence

At the center of the Hα line many spicules overlap to give the impression that the chromosphere is a fairly uniform layer. At slightly longer and slightly shorter wavelengths, however, individual spicules are more readily seen. This sequence, taken at intervals of 0.05 nm, was made on 14 October 1970.

## 3. Structure of spicules

At the boundaries of supergranular cells (**A**) material from the solar interior rises, spreads out, then descends, just as it does in the ordinary small-scale granules. The direction of flow of material in these cells is illustrated (**B**)—the "+" signs indicate rising gas, the arrows indicate horizontal radial flow, and the

10,000 km as the solar atmosphere thins out and expands over these zones. Spicules reach greater heights over the poles, and giant spicules ("macrospicules") have been found in these regions. The macrospicules contain chromospheric material at temperatures in the region of 50,000 K and rise to heights in excess of 40,000 km. They endure typically for five to ten times the average lifetime of ordinary spicules (in other words for about 40 to 50 min).

It is thought that differential motion between adjacent supergranular cells is responsible for diffusing part of the localized magnetic fields towards the polar regions. North and south polarities show a net overall tendency to collect preferentially at opposite poles. In this way the impression is created that the Sun has a general magnetic field at the poles, broadly similar to the terrestrial field, and having a strength of about 5 or 6 G (compared to 0.6 G for the Earth); indeed, for a long time astronomers were convinced that this was the case. However, it now appears that the Sun does not have an overall "dipole" field like the Earth (the sort of field that would result from a bar magnet being buried inside its globe), but only a complex array of localized surface fields (*see* pages 70–75).

Other features of the chromosphere, visible in monochromatic light, include the "fibrils", short-lived (10 to 20 min) essentially horizontal strands of gas, typically 10,000 km long and 1,000 to 2,000 km wide, which show up as dark absorption features. The pattern of fibrils is dictated by the local magnetic fields, this being particularly obvious in and around active regions. The chromosphere in monochromatic light is seen to be the seat of many aspects of solar activity, including flares, prominences, filaments, and "plages" (bright patches at higher temperatures than their surroundings).

"−" signs indicate downflow. The downflow tends to be more strongly concentrated into funnels where cells meet; the direction of flow at such a boundary is indicated (**C**). The horizontal flow of the solar plasma sweeps magnetic field lines to the edges of cells, and from these regions the spicules arise. This pattern determines the structure of the chromospheric network.

## 4. Macrospicules

Macrospicules resemble spicules but are considerably larger; they are found in the polar regions. This sequence of ultraviolet brightness contours is based on readings made on 11 December 1973. It shows at intervals of a few minutes how macrospicules rise and fall, reaching heights of about 40,000 km, some four times greater than ordinary spicules.

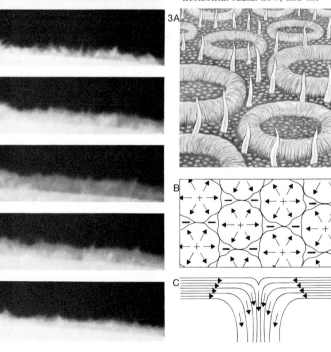

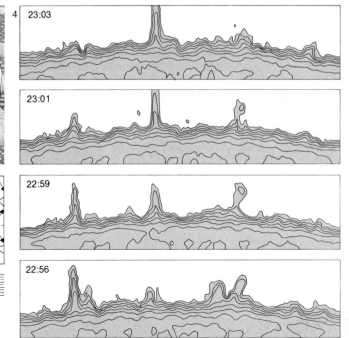

# The Chromosphere II

During an eclipse, just as the photosphere disappears from view behind the Moon, the spectrum of the chromosphere flashes into view; because of the suddenness of its appearance, this phenomenon is known as the "flash spectrum" and lasts for only a few seconds before the chromospheric light is cut off. It was first observed by the American astronomer Charles Young in December 1870.

The visible spectrum of the chromosphere consists of a very weak continuum upon which is superimposed many bright emission lines; about 3,500 lines in all have been identified. It is weak because the chromospheric gas is highly transparent to most visible wavelengths (and, as a poor absorber, it is also a poor emitter). When the solar disc is looked at directly it is impossible to see the chromospheric emission lines against the intensely luminous photosphere. Instead, the pattern of dark absorption lines is seen, arising from the absorption that takes place in the photosphere and chromosphere. When, as at a total eclipse, chromospheric gas can be seen at the limb without the background photospheric light, its lines show up clearly in emission.

Many of the lines in the chromospheric spectrum coincide with familiar absorption lines in the photospheric spectrum. $H_\alpha$ at a wavelength of 656.3 nm in the red part of the spectrum is particularly prominent, and gives the chromosphere its characteristic color. Other lines are not present in the photospheric spectrum, such as, for example, the lines of neutral and ionized

helium, and of some ionized metals. The presence of these lines indicates that some parts of the chromosphere are at appreciably higher temperatures than the photosphere. Temperature and pressure are crucial factors in determining which lines are present, and in influencing their appearance (broad, narrow, faint, strong) (*see* pages 26–27). Since both temperature and pressure change rapidly with altitude, each level displays its own characteristic spectrum. Thus, different levels in the solar atmosphere may be sampled by examining a strong Fraunhofer line and looking at wavelengths a little longer or shorter than the center of the line (*see* page 28). By selecting lines produced at different temperatures, typical of different altitudes, it is also possible to scan through the solar atmosphere and to build up a picture of its vertical structure.

For example, light emitted in the ultraviolet by singly ionized helium (HeII) at a wavelength of 30.4 nm comes predominantly from a layer with a temperature in a region around 80,000 to 90,000 K. This layer is significantly higher than the level from which most of the $H_\alpha$ radiation is emitted. Triply ionized oxygen (OIV) at temperatures in excess of 100,000 K would radiate from even higher levels of the atmosphere.

It is found that the temperature *increases* very rapidly from a minimum of about 4,200 K in the lower reaches of the chromosphere to some 500,000 K towards the top of the transition zone, the average rate of increase being about 200 K km$^{-1}$. The density

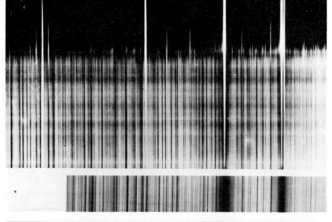

1

**1. Flash spectrum**
The flash spectrum of the chromosphere is shown with the photospheric spectrum (below) for comparison. During a total solar eclipse the normal photospheric spectrum—a continuous spectrum with dark absorption lines—is visible until the last sliver of the solar disc disappears behind the Moon. At that instant, the emission spectrum of the chromosphere, normally too faint to be seen against that of the photosphere, flashes into view for a few seconds until the chromosphere, too, is obscured by the Moon. Most of the emission lines have the same wavelengths as the dark lines in the photospheric spectrum.

**2. Layers of the chromosphere**
These spectroheliograms taken on 11 September 1961 show the different appearance of the chromosphere at four different wavelengths: H$\alpha$ (**A**), H$\alpha$ + 0.035 nm (**B**), H$\alpha$ + 0.07 nm (**C**), and the K-line of calcium (**D**). At the center of the H$\alpha$ line (**A**), areas of solar activity such as dark filaments and bright plages (also revealed in calcium K) are seen. The filaments curve towards the sunspots following the neutral line (*see* page 37). At slightly longer wavelengths (**B** and **C**) the filaments and plages fade from view. Structures at slightly higher levels of the chromosphere are shown up, since these features emit light over a relatively broad

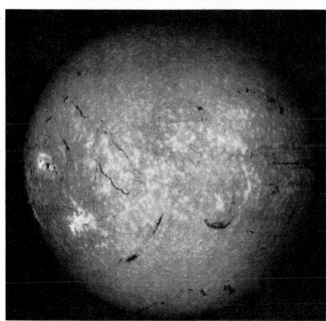

2A

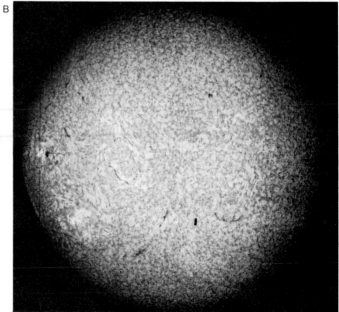

B

however *drops* equally dramatically from more than $10^{21}$ particles per cubic meter ($10^{21}\,\mathrm{m}^{-3}$) to about $10^{15}\,\mathrm{m}^{-3}$ (from about $10^{-5}$ to about $10^{-10}$ times the density of air at sea level).

The temperature in the solar atmosphere, then, drops from about 6,000 K at the base of the photosphere to the temperature minimum of some 4,200 K at an altitude of about 500 km, and thereafter increases sharply to well in excess of $10^6$ K in the lower corona. Why should this be so? According to the laws of thermodynamics, heat cannot flow from a cooler body to a hotter. It should not be possible for radiation emanating from a relatively cool body (the photosphere) to raise surrounding (coronal) material to a higher temperature than that of the emitting body. Solar energy is generated deep in the solar core and passes out through the Sun to be radiated from the photosphere at a temperature of about 6,000 K. How can the relatively low temperature of the photosphere be reconciled with the very much higher temperature of the chromosphere and corona?

In fact, the chromosphere and corona are *not* heated directly by radiation from the photosphere; the solar atmosphere above the photosphere is highly transparent to the optical and infrared radiations which comprise most of the radiation emitted from the photosphere, and as a result these radiations do not transfer significant quantities of energy to the chromosphere and corona. A popular view is that energy is transferred to the chromosphere (and to the corona) primarily by *mechanical* means: bulk kinetic energy is transmitted to the solar atmosphere and deposited there in the form of heat. What probably happens is that low frequency sound waves generated in the turbulent convective zone (rather like the rumble of thunder) move out through the photosphere and chromosphere, and the rapid decrease in density causes these waves to accelerate rapidly to become shock waves, rather like those generated by a supersonic aircraft. These shocks, moving through the chromosphere and corona, induce energetic collisions between particles with consequent heating effects.

Energy is also believed to be transmitted through the magnetic field in the form of "magnetohydrodynamic" (MHD) waves, which cause charged particles to oscillate violently. The resulting collisions contribute to raising the temperature. Another contributory effect is the downward conduction of heat from the corona, although this mechanism begs the question of why the corona is itself hotter than the chromosphere. Other mechanisms which may have a role to play include the effects of infalling material (most of which would have been hurled upwards prior to falling back), and magnetic effects, whereby currents flowing in the solar magnetic fields experience a degree of resistance and so dissipate heat in a fashion analogous to an electric bar heater. However, there is at present no completely satisfactory account of the mechanism of chromospheric heating.

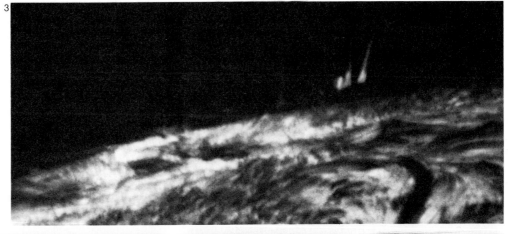

3

waveband; at $H\alpha + 0.07\,\mathrm{nm}$, for example, practically all that can be seen is the pattern of spicules outlining the chromospheric network. At the same time, however, the umbra and penumbra of the sunspots (which exist at deeper levels) can also be seen clearly, because the chromosphere is less opaque in the "wings" of the line than it is at the center of the $H\alpha$ line. A band of plages appears concentrated near the equator (**D**).

**3. Solar limb**
The lower portions of these small prominences (*see* pages 46–48) are obscured by layers of the chromosphere, here seen edge-on in $H\alpha$ light.

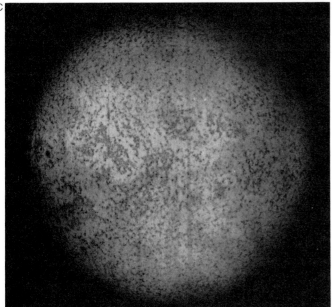

C

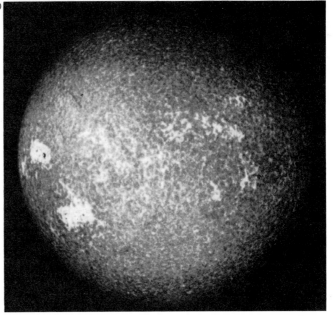

D

# Sunspots I

The existence of dark "sunspots" on the photosphere provides the most obvious evidence of activity on the Sun. Sunspots were first observed telescopically in late 1610. Although occasionally seen by naked-eye observers over the past two millenia, the fact that they were phenomena actually on the face of the Sun itself was not recognized until the era of the telescope began in the early years of the seventeenth century with the observations of Galileo, Scheiner and their contemporaries.

## Characteristics of sunspots

A typical sunspot consists of a darker central region—the "umbra"—surrounded by a lighter "penumbra". The diameter of the penumbra is, on average, about 2.5 times that of the umbra, and the penumbra can account for as much as 80 percent of the total sunspot area. The penumbra is made up of a pattern of lighter and darker filaments spreading out approximately radially from the umbra. Umbra and penumbra appear dark by contrast with the brighter photosphere because they are cooler than the average photospheric temperature; the central umbra has a temperature of around 4,000 K compared to about 5,600 K for the penumbra and about 6,000 K for the photospheric granules. Since the amount of radiation emitted by a hot body is proportional to the fourth power of its effective temperature, the umbra emits only about 30 percent of the light emitted by an equal area of photosphere, and the penumbra has a brightness of about 70 percent of the photospheric value. The darkness of a sunspot is merely an effect of contrast; if a typical sunspot, with an umbra about the size of the Earth, could be seen in isolation at the same distance as the Sun, it would shine about 50 times more brightly than the full Moon.

Observations of the Doppler effect in relatively weak spectral lines reveal a near radial flow of material outward from the umbra at speeds in the range 1 to 2 km s$^{-1}$; this is known as the "Evershed effect". However, measurements made in stronger lines—which provide information about conditions higher in the solar atmosphere—reveal the opposite effect, an inflow of material. It would seem, then, that there is a circulatory motion of gas that flows out through the penumbra, rises up vertically, then flows in a curve back down to the umbra.

Sunspots range in size from tiny pores about the size of individual granules (about 1,000 km in diameter), which appear as dark spots within penumbras, to complex structures several tens of thousands of kilometers in diameter, covering areas of up to some 10$^9$ km$^2$. A large group of sunspots may extend over a distance in excess of 100,000 km.

Sunspots generally occur in pairs or groups, isolated spots being relatively infrequent. Magnetograph observations clearly reveal that sunspots are the seats of intense magnetic fields, a typical sunspot group consisting of two spots of opposite magnetic polarity, one "north" and the other "south". The leading spot of a pair, in terms of the direction of solar rotation, is referred to as the "preceding" or p-spot while the other is known as the "following" or f-spot. With very infrequent exceptions, all the preceding spots in bipolar groups in one hemisphere have the same magnetic polarity, while all the preceding spots in the other hemisphere have the

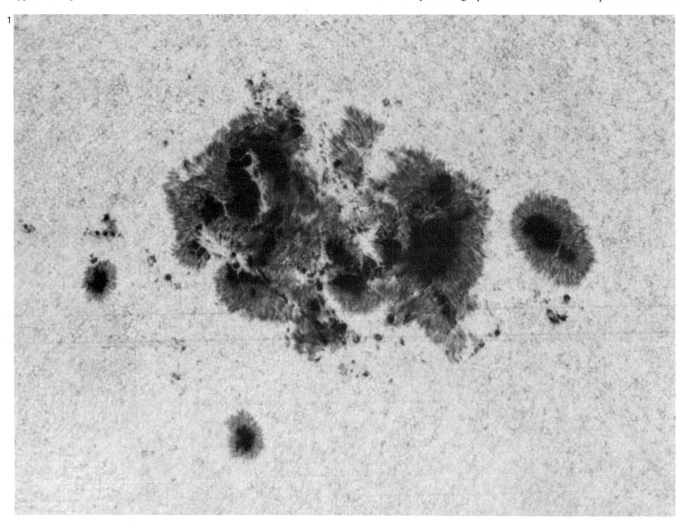

opposite polarity; similarly the following spots on either side of the equator will have opposite polarities.

The area of the solar surface covered by a sunspot group is often described in terms of millionths of the visible disc. In these terms, for example, a sunspot approximately equal in size to the Earth, would have an area of about 100 millionths. In principle, spots larger than about 500 millionths can be seen under ideal conditions by the naked eye; these comprise about 5 percent of the total.

### Magnetic fields in sunspots

The vertical magnetic fields in the umbra of a spot are usually in the range 2,000–4,000 G, up to 10,000 times stronger than the field at the Earth's surface which ranges between about 0.3 G at the equator to a maximum of about 0.7 G at a pole. The pattern of magnetic field lines between spots of opposite polarity in a bipolar group is readily apparent in spectroheliograms taken in, for example, $H_\alpha$ or the K line of ionized calcium; the pattern of fibrils, aligned along the magnetic lines of force, closely resembles the pattern obtained when iron filings are scattered on a sheet of paper placed above an ordinary bar magnet (the lines of force delineating the direction of the magnetic field around the magnet loops, which curve round from the north magnetic pole to the south).

Solar material is highly ionized, the ions and electrons having electrical charge. A charged particle can move along a magnetic line of force without experiencing resistance, but if it tries to cross the lines a secondary magnetic field is induced. The resulting force opposes the direction of motion of the particle. Thus the flow of ionized material tends to take place along the lines of force of the local magnetic field. The high electrical conductivity of solar material, resulting from its high level of ionization, ensures that matter and magnetic fields in the Sun are thus closely coupled together; the fields are said to be "frozen in" to the matter. This implies that where the energy of the field is dominant, the flow of matter will be governed by the configuration of the local magnetic field, but where the kinetic energy of the matter is greater than the magnetic energy of the field, the field lines will be distorted and will follow the mass motions. Solar magnetism is, in fact, the controlling factor in a wide variety of solar phenomena.

The sunspot groups are divided on the basis of their magnetic properties into three principal classes, as follows:

α: *unipolar* groups—single spots, or groups of spots having the same magnetic polarity;

β: *bipolar* groups, in which the p- and f-spots are of opposite magnetic polarity;

γ: *complex* groups, in which many spots of each magnetic polarity are jumbled together.

The regions of opposite polarity in a group are separated by a "neutral line" (or "line of inversion") where the vertical component of the magnetic field is zero. Dark absorbing filaments of material—which are prominences seen against the disc—are frequently found lying near neutral lines, and the violent explosive release of magnetic energy that produces flares also often occurs around these boundaries. Magnetograms show up clearly the regions of opposite polarity (*see* diagram 2B).

**1. Sunspot group**
The large group of 17 May 1951 displays a complex structure: the penumbral structure of a typical spot shows up particularly well in the detached spot to the left of the main group; the penumbra is seen to be made up of lighter and darker filaments spreading out approximately radially from the umbra.

**2. Magnetic structure**
This sequence of photographs shows the intimate link between sunspot groups, photospheric magnetic fields and the structure of the chromosphere in $H\alpha$. Sunspot groups taken at a wavelength 0.4 nm longer than $H\alpha$ are shown (**A**). The magnetogram (**B**) shows regions of positive magnetic polarity as bright patches and regions of negative polarity as dark patches, revealing the bipolar nature of sunspot groups. The influence of the magnetic field on the chromospheric structure at $H\alpha$ is shown (**C**).

**3. Bipolar sunspot group**
This photograph taken in the light of $H\alpha$ at a wavelength of 656.3 nm clearly outlines the magnetic field structure of a bipolar group. The dark filaments and fibrils follow the lines of force between the two major spots of opposite polarity, giving a pattern reminiscent of that obtained when iron filings are scattered on a sheet of paper placed over a bar magnet. The bright emission is a flare, a sudden release of energy often associated with complex sunspot groups (*see* pages 66–69).

2A 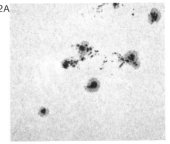 B 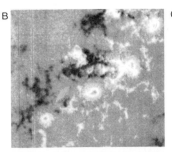 C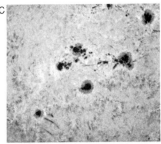

3

# Sunspots II

Various systems have been devised for the classification of sunspots; the one shown on this page (*see* diagram 2) is the "Zurich Spot Classification", which divides sunspots into nine classes designated by the letters A, B, C, D, E, F, G, H and J. Essentially the system relates to the evolutionary stages through which groups of spots pass, although not all spots go through the complete sequence. Starting as a small cluster (type A) or as a bipolar group (B), spots grow rapidly and within a period of 8 to 10 days they arrive at their maximum area (F) after passing through the intermediate stages (C, D and E). Decay takes slightly longer, so that the group spends most of its life in the final stage of the classification (G to J). A group of moderate size, however, might follow an abbreviated sequence of development, while a very small group might not develop beyond the first stage. Associated activity, such as flares, tends to reach a maximum fairly early in the evolution of the group, typically during stages D, E and F. The total lifetime of a large group may be as long as several weeks; the average is less than two weeks. Lifetime is approximately related to the area of the sunspot group: if the area is expressed in millionths of the visible disc, then a rough guide to the lifetime of the group in days is given by dividing the area by 10. Thus a small spot of 10 millionths would be likely to last only a day, while a spot 100 millionths in area would be expected to survive for about 10 days.

## Sunspots and solar rotation

The Sun rotates on its axis, but unlike a solid body different parts of the surface rotate in different periods of time. The mean rotation period of the Sun—or of its surface at least—is 25.380 days. However, from Earth it appears to be rotating more slowly, because the Earth is moving around the Sun in the same direction as the Sun is spinning on its axis; the apparent mean rotation period (known as the "synodic period") is 27.275 days.

The first means of deducing the period of solar rotation was provided by the motion of sunspots across the solar disc. Observations made on successive days reveal that sunspots change their positions on the visible disc, marching steadily across from east to west (from the eastern to the western limb of the Sun as seen in the sky) as they are carried around by the rotation of the Sun. The apparent paths taken by sunspots across the disc vary throughout the year because the solar equator is tilted to the plane of the ecliptic by an angle of about $7°.25$.

The position of features such as sunspots on the photosphere is specified by giving their "heliographic" latitudes and longitudes. These are coordinates essentially similar to latitude and longitude on Earth. Heliographic latitude is measured north or south of the solar equator (taking values from $0°$ to $\pm 90°$), while heliographic longitude is measured along or parallel to the solar equator from a standard meridian (an imaginary circle perpendicular to the equator and passing through the poles of the Sun). This meridian is defined as having crossed the center of the visible disc on 1 January 1854 at 12.00 U.T. and is assumed to have been rotating with a uniform period of 25.38 days ever since. The system was devised by Richard Carrington (*see* page 18) and does not relate to any specific visible feature of the solar disc, unlike terrestrial longitude, which operates with reference to the meridian passing through Greenwich; instead heliographic longitude is based on an imaginary meridian which revolves around the center of the Sun at the rate which Carrington assigned to the mean rotation of the Sun.

On, or about, 6 June and 6 December each year the heliographic equator is seen edge-on from the Earth, and sunspots follow

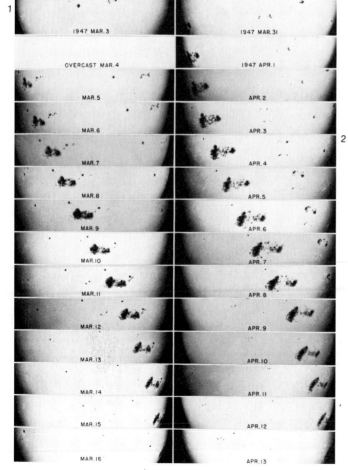

**1. Solar rotation**
This sequence shows the giant sunspot group of 1947. After passing round the far side of the Sun the group reappears to make a second crossing. This group, the largest ever recorded, reached a maximum area of $18 \times 10^9$ km².

**2. Zurich classification scheme**
This system of classification was developed by M. Waldmeier. Additional information may be added by means of a three-letter code (Zpc). The first letter denotes the basic Zurich class (A–J); the second denotes the penumbral type (large, small, etc.) of the largest spot in the group, and the third denotes the degree of compactness of the distribution of the spots.

straight-line paths across the disc. At other times of the year sunspots follow slightly curved paths depending upon which pole of the Sun is tilted towards the Earth. About 7 March the south heliographic pole is tilted towards the Earth by the maximum amount, and the solar equator passes 7°.25 north of the center of the visible disc; around 8 September the north pole is inclined towards Earth by the maximum amount and the solar equator passes 7°.25 south of center. The angle which the north–south axis of the Sun makes with the true north–south direction in the sky varies throughout the year because of the tilt of the solar axis and the inclination of the ecliptic to the celestial equator (*see* Table).

Moreover, the motion of sunspots at different heliographic latitudes reveals the "differential rotation" of the Sun: the photosphere rotates at different rates at different latitudes, with rotation periods ranging from about 25 days at the equator to about 27 days at latitudes ± 30°; still longer periods are found at higher latitudes. Since sunspots are seldom found more than about 40° from the equator, the rotation periods of higher latitude regions are more difficult to determine.

However, the rotation rate of the photosphere may also be deduced from the Doppler effect (*see* page 27) in the wavelengths of spectral lines originating in the approaching and receding limbs at different latitudes. Light from the approaching limb will be blue-shifted, while light from the receding limb will be red-shifted. By measuring the Doppler shift in this way it has been shown that the rotation period ranges from 26 days at the equator to 37 days at the poles. The sunspots appear to rotate faster than the general photospheric background by about 4 to 5 percent. (These rotation periods are *sidereal* periods, true rotation periods with respect to the background stars. Since the Earth revolves around the Sun in

the same direction as the Sun rotates on its axis, the apparent rotation rates are slower; the apparent or synodic periods range from 28 days at the equator to 41 days at the poles.)

**The Wilson effect**
When a sunspot is carried across the Sun's disc to the limb it is possible to view the spot almost edge-on. The umbra is often seen to be displaced towards the center of the disc in such a way as to give the impression that a sunspot is a saucer-shaped depression, the apparent length of the center of an average spot being about 800 km. This is known as the Wilson effect, after the Scottish astronomer A. Wilson, who first noted it in 1769. However, because the gas in the umbra is less dense than in the surrounding photosphere, it is more transparent (the opacity is lower) and it is possible to see to a greater depth; consequently, the apparent depression of the umbra is probably an illusion, since at the umbra it is possible to see through the solar material to a greater depth.

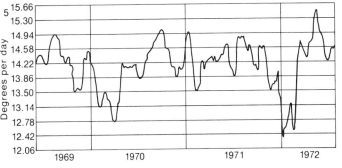

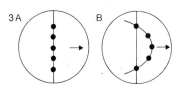

**3. Differential rotation**
The rotation period of the photosphere increases with increasing latitude. In the idealized situation shown here, if a row of sunspots lay along the central meridian of the Sun (**A**) then, after one rotation, the spots would be spread out (**B**).

| 5 Jan | 8 Feb | 7 March | 8 April | 8 May | 5 June |

| 7 July | 13 Aug | 8 Sept | 11 Oct | 9 Nov | 7 Dec |

**4. Apparent sunspot motions**
In the course of a year the angle between true north and the Sun's axis takes values between 0° and 26°.3 east or west (*see* Table below) because of the combined effects of the tilts of the solar and terrestrial equators to the plane of the Earth's orbit.

**5. Rotation rate of the Sun**
In addition to the variations in the Sun's rate of rotation at different latitudes (*see* diagram 3), there are also considerable fluctuations about the mean equatorial value of 13°.86 per day. The reason for these changes in the rotation rate are not clear.

**Variation of the position angle of the Sun's axis (north pole)**

| 5 Jan | 0° | 8 April | 26°.3 | 7 July | 0° | 11 Oct | 26°.3 |
|---|---|---|---|---|---|---|---|
| 16 Jan | 5° | 26 April | 25° | 19 July | 5° | 30 Oct | 25° |
| 27 Jan | 10° | 8 May | 23° | 30 July | 10° | 9 Nov | 23° |
| 8 Feb | 14° | 19 May | 20° | 13 Aug | 14° | 20 Nov | 20° |
| 23 Feb | 20° | 5 June | 14° | 28 Aug | 20° | 7 Dec | 14° |
| 7 March | 23° | 15 June | 10° | 8 Sept | 23° | 16 Dec | 10° |
| 18 March | 25° | 26 June | 5° | 21 Sept | 25° | 26 Dec | 5° |

**Variation of the heliographic latitude of the center of the Sun's disc**

| 1 Jan | 3°.0 S | 21 Apr | 5°.0 S | 1 July | 3°.0 N | 25 Oct | 5°.0 N |
|---|---|---|---|---|---|---|---|
| 10 Jan | 4°.0 S | 2 May | 4°.0 S | 11 July | 4°.0 N | 4 Nov | 4°.0 N |
| 19 Jan | 5°.0 S | 11 May | 3°.0 S | 21 July | 5°.0 N | 13 Nov | 3°.0 N |
| 1 Feb | 6°.0 S | 20 May | 2°.0 S | 24 Aug | 6°.0 N | 21 Nov | 2°.0 N |
| 18 Feb | 7°.0 S | 28 May | 1°.0 S | 24 Aug | 7°.0 N | 24 Nov | 1°.0 N |
| 7 Feb | 7°.3 S | 6 June | 0° | 9 Sept | 7°.3 N | 7 Dec | 0° |
| 21 Mar | 7°.0 S | 14 June | 1°.0 N | 22 Sept | 7°.0 N | 15 Dec | 1°.0 S |
| 8 Apr | 6°.0 S | 22 June | 2°.0 N | 12 Oct | 6°.0 N | 23 Dec | 2°.0 S |

**6. The Wilson effect**
A sunspot looks like a saucer-shaped depression when seen close to the limb. This sequence shows a small, almost circular spot having a symmetrical umbra and penumbra. The picture taken close to the limb shows that the apparent width of the umbra on the side further from the limb is less than the apparent width on the other side. Although this is consistent with the idea of a sunspot being a depression, the effect may be an illusion.

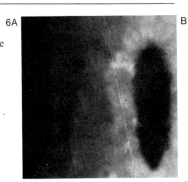

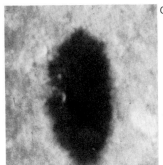

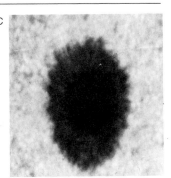

# The Sunspot Cycle

The number of sunspots visible on the solar disc varies in a periodic way. This phenomenon was discovered by Heinrich Schwabe, who published his analysis of 17 years of observational data in 1843. At a maximum in the cycle there may be 100 or more spots on the disc, but at a minimum there are very few spots, and several weeks may pass without any at all being seen.

Schwabe's initial announcement was of a sunspot cycle of about 10 years, but later analysis by R. Wolf led to the more accurate value of 11 years, which is quoted conventionally today. However, over the past 50 years or so the period between successive maxima has averaged out at about 10.4 years. Since records began, the duration of individual cycles has varied from about 7 to about 17 years. The "11-year" cycle is readily apparent in a graph of sunspot numbers, which also shows some evidence for a longer term modulation of the height of maximum over a period of about 80 years. In 1893 E. W. Maunder of the Royal Greenwich Observatory concluded from his study of old solar records that for a period of about 70 years from about 1645 to about 1715 sunspot activity virtually vanished. Recent investigations have largely confirmed these conclusions, and have also shown that similar bouts of inactivity appear to have occurred in the more distant past. During the "Maunder Minimum" from 1645 to 1715 the appearance of a sunspot was regarded as a most noteworthy event; John Flamsteed, the first Astronomer Royal, at one stage scanned the Sun for fully seven years between sightings of a spot. Although absence of evidence of spots is not necessarily evidence of their absence, it seems unlikely that astronomers should not have maintained observations of the Sun. Sunspots were first studied telescopically in about 1610, but the 11-year cycle was not discovered until 1843: if the cycle had been behaving as regularly as it now does, it seems likely that it would have been noticed before Schwabe's time.

Another feature of the sunspot cycle is the cyclic change in the mean latitudes at which spots occur. At the beginning of a new cycle, spots tend to appear at latitudes of between 30° and 40° north and south of the equator, but as the cycle progresses they occur at successively lower latitudes. At maximum, spots tend to be found in zones at about ±15°, and by minimum the mean latitude of sunspots is between 5° and 7°; as the last spots of the old cycle are occurring in low latitudes, the first spots of the new cycle are beginning to emerge in higher latitudes once more. The progression of regions of spot activity towards the equator during each cycle is known as "Spörer's law"—the effect having been discovered in 1858 by R. Carrington and then investigated in more detail by G. Spörer—and is graphically represented in the so-called "butterfly diagram" (see diagram 3). In sunspot groups the preceding spot is usually found to lie at a slightly lower latitude than the following one and, whereas near the equator the line joining the two principal spots in a group may be inclined to the equator by as little as 1°, at high latitudes the inclination may be as great as 20°.

## Measurement of sunspot activity

The level of sunspot activity is assessed by the Zurich (or "Wolf") relative sunspot number, R, which is defined somewhat arbitrarily as $R = k(f + 10G)$, where f is the total number of spots, G is the number of groups and k is a factor that depends upon the idiosyncracies of the individual observer and his telescope. For example, if there were three groups on the Sun, containing, respectively, 3, 6 and 10 spots, f would be 19, and G would be 3; assuming, for simplicity, k to be 1, the value of R would be $19 + (3 \times 10) = 49$. This means of measuring sunspot activity is not altogether satisfactory: for example, five groups of two spots each would give R a value of 50, compared to 49 for 19 spots in the

**1. Solar minima and maxima**
This pair of photoheliograms illustrates the contrast in the appearance of the Sun at solar minimum (**A**) and at solar maximum (**B**). The former was taken at 8 hr 32 min UT on 5 September 1974, near the time of a sunspot minimum; only a few small isolated spots may be seen. At minimum the solar disc may appear completely blank. The right-hand photograph was taken at 9 hr 46 min UT on 12 November 1979, during one of the more active solar maxima of the twentieth century. Large sunspot groups are seen to be confined to two bands on opposite sides of the equator.

**2. Historical record**
The sunspot cycle is displayed over a period of nearly four centuries in this graph of mean annual values of the Wolf Relative Sunspot Number. The basic 11-year cycle is readily apparent in the more recent data, although it is clear that there are variations in the lengths of the cycles and in the levels of activity at successive maxima. The period from 1645 to 1715 is the Maunder Minimum, when, it would seem, sunspots were almost completely absent. In addition to the 11-year cycle, there are indications of a longer-term modulation of the heights of successive maxima with a period of, perhaps, 80 years.

1A

B

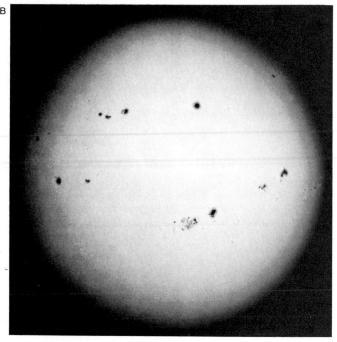

previous example; nevertheless, it has the merit that detailed records of R date back to 1848, when Rudolph Wolf introduced the system; Wolf himself by examining old observations, extended the sunspot record back to the time of Galileo and Scheiner. (A more precise measure of sunspot activity can be given by considering the total *areas* of sunspots.)

## Reversal of polarity

With very few exceptions, if all the p-spots in the northern hemisphere happen to have north magnetic polarity, all the p-spots in the southern hemisphere will have south polarity. At the end of each 11-year cycle the pattern of spot polarities reverses. Continuing the previous example, all the new spot pairs appearing around latitude 30° north would have south magnetic poles associated with their p-spots, while the p-spots of the corresponding new groups in the southern hemisphere would have north magnetic polarity. Because of the polarity reversal at the end of each 11-year cycle, 22 years elapse before the Sun returns completely to its original overall pattern.

Associated with this phenomenon is the periodic reversal of the weak "polar field". Although the Sun does not have an overall dipole field like that of the Earth, it has instead a field made up of large numbers of localized flux elements spread through the outer layers of its globe. Nevertheless there are large areas of net polarity at the polar regions, with the polarity at the north heliographic pole (usually) opposite to that at the south heliographic pole. Recent estimates of the net polar field are around 5 or 6 G at solar minimum. The polar magnetic polarities tend to reverse about a year or so after the sunspot maximum, in other words (about half-way through the cycle) since the rise to maximum activity tends to be more rapid than the decline to minimum (the average rise time is

4.6 years compared to the average decline of 6.7 years). The change-over at opposite poles generally does not occur simultaneously, and on occasions one polarity can change as much as a year or two before the other. For example, the south magnetic pole became a north magnetic pole in 1957, but the north magnetic pole did not become a south magnetic pole until the end of 1958; thus, for a time, the Sun may have two north or two south magnetic poles.

**Zurich yearly means of daily relative sunspot numbers**

| Year | No. | Year | No. | Year | No. | Year | No. | Year | No. |
|------|-----|------|-----|------|-----|------|-----|------|-----|
| 1851 | 64.5 | 1877 | 12.3 | 1903 | 24.4 | 1929 | 65.0 | 1955 | 38.0 |
| 1852 | 54.2 | 1878 | 3.4 | 1904 | 42.0 | 1930 | 35.7 | 1956 | 141.7 |
| 1853 | 39.0 | 1879 | 6.0 | 1905 | 63.5 | 1931 | 21.2 | 1957 | 190.2 |
| 1854 | 20.6 | 1880 | 32.3 | 1906 | 53.8 | 1932 | 11.1 | 1958 | 184.6 |
| 1855 | 6.7 | 1881 | 54.3 | 1907 | 62.0 | 1933 | 5.6 | 1959 | 159.0 |
| 1856 | 4.3 | 1882 | 59.7 | 1908 | 48.5 | 1934 | 8.7 | 1960 | 112.3 |
| 1857 | 22.8 | 1883 | 63.7 | 1909 | 43.9 | 1935 | 36.0 | 1961 | 53.9 |
| 1858 | 54.8 | 1884 | 63.5 | 1910 | 18.6 | 1936 | 79.7 | 1962 | 37.5 |
| 1859 | 93.8 | 1885 | 52.2 | 1911 | 5.7 | 1937 | 114.4 | 1963 | 27.9 |
| 1860 | 95.7 | 1886 | 25.4 | 1912 | 3.6 | 1938 | 109.6 | 1964 | 10.2 |
| 1861 | 77.2 | 1887 | 13.1 | 1913 | 1.4 | 1939 | 88.8 | 1965 | 15.1 |
| 1862 | 59.1 | 1888 | 6.8 | 1914 | 9.6 | 1940 | 67.8 | 1966 | 47.0 |
| 1863 | 44.0 | 1889 | 6.3 | 1915 | 47.4 | 1941 | 47.5 | 1967 | 93.8 |
| 1864 | 47.0 | 1890 | 7.1 | 1916 | 57.1 | 1942 | 30.6 | 1968 | 105.9 |
| 1865 | 30.5 | 1891 | 35.6 | 1917 | 103.9 | 1943 | 16.3 | 1969 | 105.5 |
| 1866 | 16.3 | 1892 | 73.0 | 1918 | 80.6 | 1944 | 9.6 | 1970 | 104.5 |
| 1867 | 7.3 | 1893 | 84.9 | 1919 | 63.6 | 1945 | 33.1 | 1971 | 66.6 |
| 1868 | 37.3 | 1894 | 78.0 | 1920 | 37.6 | 1946 | 92.5 | 1972 | 68.9 |
| 1869 | 73.9 | 1895 | 64.0 | 1921 | 26.1 | 1947 | 151.5 | 1973 | 38.0 |
| 1870 | 139.1 | 1896 | 41.8 | 1922 | 14.2 | 1948 | 136.2 | 1974 | 34.5 |
| 1871 | 111.2 | 1897 | 26.2 | 1923 | 5.8 | 1949 | 134.7 | 1975 | 15.5 |
| 1872 | 101.7 | 1898 | 26.7 | 1924 | 16.7 | 1950 | 83.9 | 1976 | 12.6 |
| 1873 | 66.3 | 1899 | 12.1 | 1925 | 44.3 | 1951 | 69.4 | 1977 | 27.5 |
| 1874 | 44.7 | 1900 | 9.5 | 1926 | 63.9 | 1952 | 31.5 | 1978 | 92.5 |
| 1875 | 17.1 | 1901 | 2.7 | 1927 | 69.0 | 1953 | 13.9 | 1979 | 155.4 |
| 1876 | 11.3 | 1902 | 5.0 | 1928 | 77.8 | 1954 | 4.4 | 1980 | 154.6 |

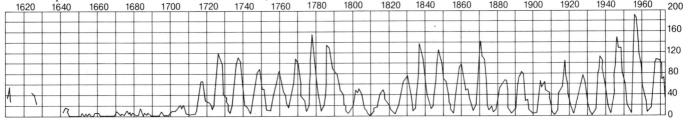

## 3. Butterfly diagram

The numbers and latitudes of sunspots have varied over 8 complete cycles between 1874 and 1976. The latitude of each spot is plotted against the dates on which it was visible. The diagram shows in dramatic form how the first spots of a new cycle appear in small numbers at relatively high latitudes. Numbers increase as the maximum is approached, and at maximum the mean latitude of sunspots is about ± 15°. At the end of the cycle a few spots are seen close to the equator. The distribution of spots on the diagram resembles the wings of a butterfly.

## 4. Sunspot areas

This graph plots the area of the visible disc covered by sunspots, expressed in millionths of the visible hemisphere. The smooth curve through the "spikes" shows the variation in mean annual sunspot areas. The variation in sunspot areas follows the same basic pattern as the variation in sunspot numbers shown in the butterfly diagram.

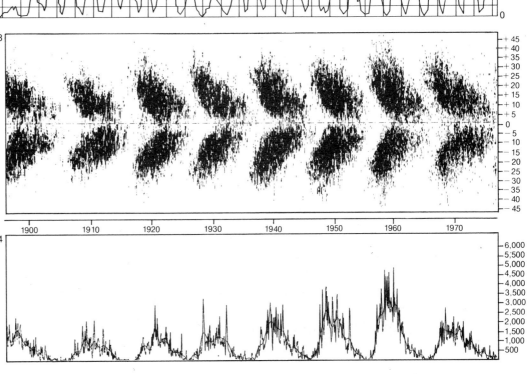

# Theories of Sunspots

There have been many theories of the nature of sunspots, ranging from the suggestion that they were planets passing in front of the Sun, to the idea that they were solar storms analogous to cyclones on Earth. Particularly fascinating was the suggestion advanced in the early nineteenth century by Sir William Herschel (discoverer of the planet Uranus and the leading observer of his day) that sunspots were holes in the fiery luminous outer layers of the Sun through which it was possible to see towards the solid surface on which, he believed, living creatures almost certainly existed.

## Leighton-Babcock model

At present there is no completely satisfactory theory of sunspots. There is, however, one general qualitative model which has many attractive features and which is generally considered to give a fair account of most aspects of sunspots and their behavior; this model was proposed in 1961 by H. W. Babcock and developed further, notably by R. Leighton.

A weak magnetic field may be imagined permeating the outer solar layers, possibly as far down as the base of the convective zone, some 200,000 km below the photosphere; this field can be represented initially by lines of force running along meridians from north to south. The high conductivity of solar material ensures that these field lines will be frozen into the matter. Because of the Sun's differential rotation, the field lines will be stretched out and wrapped around the solar globe, like elastic threads. As the lines are stretched out they are concentrated closer and closer together, amplifying the strength of the surface field; in effect the differential rotation of the Sun "winds up" the initially weak fields to much higher strengths. Recent studies show that the magnetic field lines are concentrated in thin tubes of magnetic "flux". (Flux measures the total field passing through an area perpendicular to the direction of the field; it can be regarded as the number of field lines passing perpendicularly through a given area.) The flux tubes are a few hundred kilometers in radius, and within them the strength of the field reaches values between a few hundred gauss and about 2,000 G. Hot bubbles of rising gas will carry magnetic flux with them, distorting the field lines. Such convection effects cause the flux tubes to become twisted, thus amplifying further the magnetic fields contained in them. Bundles of flux tubes become tangled together and compressed, eventually producing field strengths as great as 2,000 or 4,000 G. With increasing field strength the magnetic pressures in the tangled flux tubes become sufficiently great for them to become buoyant, and float up to the surface. Where the bundles of twisted flux tubes pierce the photosphere, the field spreads out to form a sunspot group. Individual flux tubes penetrating the surface probably correspond to spicules.

The winding-up of the field causes this process to occur first at latitudes around $\pm 40°$, for it is here that the shearing motion between adjacent zones of gas is at its greatest. As spots emerge at these relatively high latitudes the fields at lower latitudes are enhanced, and the zones in which magnetic eruptions take place migrate progressively towards the equator. This is in accordance with the observed behavior of sunspot groups. Since the field lines slope towards the equator, the preceding spot of an emerging pair should occur at a slightly lower latitude than the following one; this, again, is in qualitative agreement with observation.

As the bipolar magnetic region spreads out and declines, the magnetic flux is carried away and spread around by the formation and destruction of supergranular cells and by differential rotation. Because the f-spot is closer to the pole in each case, its polarity tends preferentially to be carried to the polar regions of the hemisphere in which it is located, building up sufficient strength by around the

## 1. Magnetogram

This magnetogram (A), obtained on 5 July 1978, shows opposite polarities as black and white patches. Complex bipolar regions are associated with large groups; the polarities of p- and f-spots are reversed on either side of the equator. The white light picture (B) shows sunspot distribution.

## 2. Magnetic field reversal

An idealized diagram shows the north polar region with a positive net polarity at an early stage of a cycle (A); the south polar region is negative. In the northern hemisphere, p-spots are positive. Around solar maximum, the polar fields are neutral (B), and in the next cycle they are reversed (C).

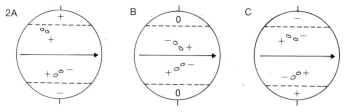

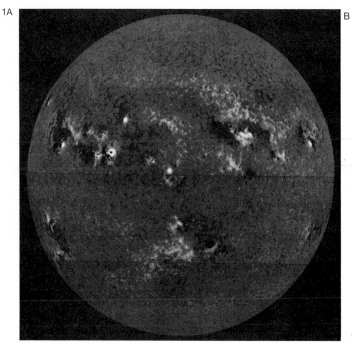

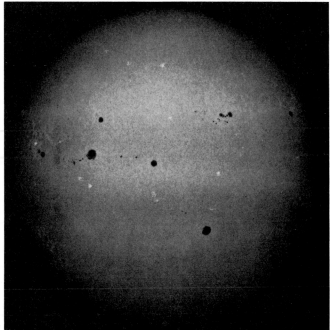

time of sunspot maximum to reverse the existing polarity there.

It is thought that as the polarity in a polar region changes, it begins to change the slope of the field lines in the locality; as the strength of the new polar field builds up (a result of the continued diffusion of f-spot polarity) so the zone in which slope reversal occurs moves closer to the equator. When the slope of the field lines changes from downwards (in the direction of solar rotation) to upwards, the effect of differential rotation is to unwind the field lines, rather than winding them up. The concentrated fields required for the production of sunspot groups decline and activity diminishes. Eventually a stage is reached where, once again, the lines of force run approximately from north to south and the winding-up process begins again. Since the field is in the opposite direction, the polarities of p- and f-spots will be reversed compared to the previous cycle. As the reversal begins to produce spots in higher latitudes, the residual field from the old cycle still gives rise to some spots of the old polarity close to the equator.

In this way, the Sun is thought to wind up and relax its magnetic fields in a periodic fashion. Magnetic energy built up during a cycle is released in various forms of solar activity, the most violent being the solar flare (*see* pages 66–69). Although this model has much to commend it, solar activity and its cyclic behavior is still far from being fully understood. Indeed, one key question is as yet unanswered: where does the magnetic field originate?

The most popular explanation is the "dynamo model", whereby the circulation of electrical currents (comprising ions and electrons) in the convective zone is responsible for generating the fields. A moving electrical charge generates a magnetic field: with powerful convective forces driving the highly conductive solar material, substantial fields can be generated and sustained. Similar processes in the hot liquid metallic core of the Earth are believed to be responsible for sustaining the terrestrial magnetic field.

According to E. N. Parker, the magnetic fields must be generated in the lower reaches of the convective zone (about 200,000 km below the photosphere), where they cannot be much stronger than 100 G otherwise they would rise to the surface and escape too rapidly to fit in with the solar cycle. On the other hand they cannot be much less than 100 G if the observed outflow of magnetic flux during a cycle is to be maintained. However, there remain difficulties in accounting for the concentration of the field into various flux tubes, and for the way in which bundles of flux tubes can be squeezed together to give the observed field strengths in sunspot umbras, despite their natural tendency to repel each other. Powerful convective forces must be at work to achieve this result. Furthermore, there is no convincing theoretical explanation of the actual 22-year period of the sunspot cycle.

## Why are sunspots cool and dark?

Although sunspots have been studied for centuries, there is still some debate about the precise reason why they are dark. It was believed that the intense localized magnetic field in the umbra reduced the flow of hot material to that region of the photosphere; as a result less radiant energy is transported to the photosphere in that area than in its surroundings. It has been suggested, however, that this theory of sunspot darkening is not altogether satisfactory since it should lead to a build-up of heat below the spot. An alternative view suggests that the strong magnetic field actually enhances the flow of heat, but converts 75 to 80 percent of this energy into "hydromagnetic waves", which propagate through the photospheric layer without dissipating significant amounts of energy; the wave energy contributes instead to the heating of the solar atmosphere higher up above the sunspot group.

**3. Babcock-Leighton model**
According to this model the basic mechanism responsible for sunspot activity is the winding up of the solar magnetic field by the Sun's differential rotation. An idealized situation is represented (**A**) in which the north pole is taken to have negative magnetic polarity and the south pole, positive polarity. Lines of force lie along meridians in a north–south direction on and below the photosphere. The magnetic field lines are "frozen in" the solar material, and in time become distorted by the differential rotation (**B**). As the cycle progresses, the lines become stretched and wrapped round the Sun. By this means the field lines are drawn together and the field strength greatly enhanced (**C**). Field lines are brought together to form tubes of magnetic flux (**D**), which become twisted by the effects of convection. Bundles of tubes may be wound together into structures rather like ropes, this process amplifying still further the magnetic field strength. When the field strengths in the tubes or ropes becomes sufficiently great, they float to the surface, and where kinks in the bundles of tubes penetrate the photosphere, a sunspot group is formed. Below a sunspot (**E**) magnetic tubes of force are squeezed together—despite their natural tendency to push apart—by powerful convective currents.

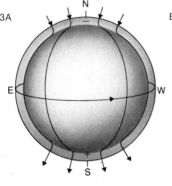

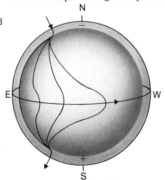

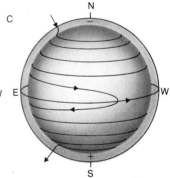

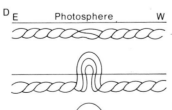

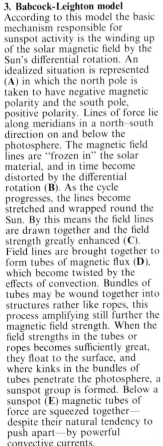

**4. Zones of activity**
The enhanced fields first reach maximum strength at latitudes of about $\pm 40°$. As a result, sunspot formation tends to occur first in these latitudes (**A**). During the winding-up processes the field lines slope towards the equator so that the p-spot of the pair tends to lie closer to the equator than does the f-spot. As strong fields from higher latitude spots weaken and spread out, the zone of sunspot activity moves closer to the equator. The distended field of an old bipolar group is shown (**B**).

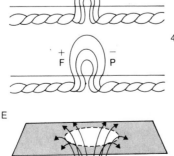

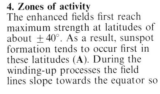

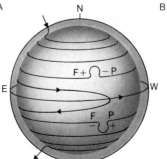

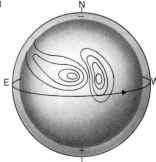

# Active Regions

Associated with sunspots are various other indications of solar activity, phenomena such as prominences and flares (*see* pages 46–48 and 66–69), faculae (*see* page 31), plages and filaments, which are all controlled by the intense localized magnetic fields that characterize "active regions".

## Plages

Named after the French word for "beaches", plages are areas of intensified brightness visible in various monochromatic lines; they look, in a way, like bright, sandy beaches against the background of the chromosphere. Plages coincide approximately in position with the photospheric faculae, and represent regions of enhanced density and temperature which are, in turn, overlaid by hotter regions of the corona; they delineate the area of enhanced magnetic field, which constitutes a complete active region. Like photospheric faculae, they appear before the emergence of sunspots and persist long after the spots themselves have faded away. Indeed, the term "chromospheric faculae" is sometimes used to describe them.

The brightness of plages is believed to be due to the increased flow of energy into the solar atmosphere which is brought about by the action of the concentrated magnetic fields. The deposition of energy transported by mechanical means is believed to be the main source of heating in the solar atmosphere (*see* pages 34–35). Thus the process responsible for heating the plages is basically the same as that which produces the bright regions that outline the chromospheric network. Although the very strong fields of 2,000 to 4,000 G found in sunspot umbrae clearly limit the flow of heat to those regions of the photosphere, the weaker overall fields in plages (which average 100 to 200 G) enhance the deposition of energy in those features and in the corona above.

Plages generally appear brighter in shorter wavelengths, which indicates that they are brighter at higher altitudes. Radio plages seen at much longer, decimetric, wavelengths are relatively high-density regions of the corona which overlay the visible plages and have brightness temperatures of about $10^6$ K. They are the main source of the S-component of solar radio emission (*see* page 28).

## Filaments

$H_\alpha$ spectroheliograms reveal the presence of dark filaments of absorbing material which may be up to, or even in excess of, 100,000 km in length. They form along the neutral line (the line dividing regions of opposite magnetic polarity in an active region— *see* page 37) and may persist for several months (6 to 10 months is not uncommon); sometimes they fade from view and then re-form in more or less the same locations.

Filaments are tubes and loops of relatively concentrated matter at chromospheric temperatures lying above the sunspot groups, in the upper chromosphere and low corona. Seen against the bright background of the solar disc they show up because of their absorption of light, but seen at the limb they are about 100 times brighter than the background corona, and show up clearly as emission features; they are then known as prominences.

## Structure and evolution of an active region

Most active regions are bipolar magnetic regions (BMRs), which have approximately equal amounts of flux of the opposite magnetic polarities. Although in sunspots the average flux in the p-spot tends to be higher than that in the f-spot, the difference is ironed out in the overall field of the active region. Unipolar spot groups usually turn out to be old BMRs from which the f-spots have vanished (the p-spot, being stronger, tends to last longer); the distribution of faculae and plages in the vicinity usually has the form typical of a bipolar region, with a magnetic region of opposite polarity to the visible spot usually found in the location of the "missing" f-spot. The neutral line is picked out along parts of its length by filaments and by plage "corridors", darker lanes that lie between bright plages.

The formation of an active region is generally a quite rapid

1A

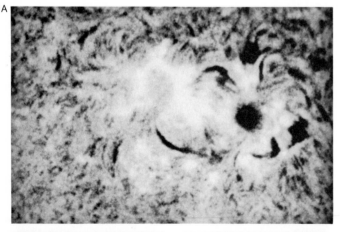

**1. Active region**
A group of sunspots, together with a plage and several dark filaments are visible in the Hα image of a typical active region (**A**). The coronal loops above this region are shown up at X-ray wavelengths (**B**), while the magnetogram (**C**) reveals the distribution of magnetic polarities (black and white representing opposite magnetic polarity). The loops in the corona can be seen arching across and joining up opposite polarities in the chromosphere below.

**2. Development of an active region**
This sequence of two sets of images was taken at the Big Bear Solar Observatory, California, in 1972. The top set shows an active region in Hα, best for revealing plages, filaments and the magnetic pattern of the fibrils, while the lower set is in Hα–0.1 nm, providing a clearer image of the sunspots. As the region develops, the neutral line dividing the preceding spots from the following spots rapidly becomes more twisted and the spots grow in size.

2A

30 July     17.35

16.18

B

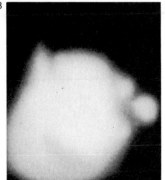

C

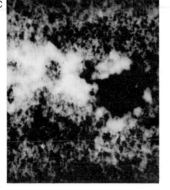

B

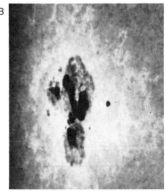

process, a typical region taking about 10 days to build up to its maximum extent. As the local magnetic field begins to increase with the emergence of magnetic flux from below the photosphere, features such as fibrils begin to line up along the field lines. A small plage will often make its appearance and thereafter, although not always, spots will begin to form. Prominence and flare activity escalates rapidly during the first couple of weeks.

Usually after about two to four weeks (unless it is a particularly large group) almost all of the sunspots will have disappeared, but the BMR itself can persist for several months, declining slowly by spreading out and breaking up as its field is eroded. The size of an active region usually determines its lifetime. While flares are normally associated with active sunspot groups, they can also occur in spotless BMRs.

## Other aspects of solar activity

X-ray pictures reveal that the face of the Sun is speckled with bright points that correspond to small short-lived active regions where sunspots are absent; they endure for between a few minutes and a few days, but typically survive for only a few hours. Although much smaller than sunspot groups they display the characteristics of bipolar magnetic regions with intense localized fields, which appear to follow the same polarity rules with regard to p- and f-regions. Bright points tend to be found at the boundaries of supergranular cells where the solar field is enhanced but, although they show a slight preference for middle latitudes, they are distributed widely over the entire solar globe, including the polar regions. At any one time there are typically about 100 bright points on the visible disc.

There is some debate as to just how much magnetic flux is contained in these ephemeral regions, but it does seem to be a significant fraction of the total that emerges from the Sun.

There is some evidence to suggest that their numbers increase as the frequency of other aspects of solar activity—sunspots, active regions, flares—declines. This may imply that the level of solar activity evens out over the cycle such that the sunspot cycle alone may not give a full account of the degree to which magnetic activity fluctuates. A tendency to increase in numbers towards the minimum of the solar cycle is also displayed by polar faculae, the numbers of which reach a minimum around solar maximum and increase thereafter. Polar faculae are small, round, bright points seen in optical wavelengths; they are typically about 2,000 km in diameter and last, on average, for only some 30 min. Recent results indicate that during the minimum of 1973–75 the total magnetic flux above latitudes $\pm 40°$ increased by about 250 percent. The increase in numbers of polar faculae—which are, after all, magnetic phenomena—may be linked to this change.

Although the most obvious aspect of the overall magnetic field of the Sun is the existence of regions of opposite polarity at opposite poles, there are other large-scale aspects. As the decaying fields from BMRs spread out and are dissipated, with a net drift of f-spot polarity to the polar region, the flux also shows a predilection to form large-scale regions, or sectors, of alternating net magnetic polarity; these zones can span a few tens of degrees to more than 100 degrees of heliographic longitude and may endure for as long as three years. The individual elements of flux change continually while the overall pattern remains essentially constant. Although of weak overall field strength, these large unipolar magnetic regions appear to have a significant effect on the behavior of the solar wind (*see* pages 76–77).

The wealth of new data from sources such as Skylab and the Solar Maximum Mission has shown clearly that all forms of solar activity, as well as the structures of the chromosphere and the corona (*see* pages 70–73), are controlled by solar magnetic fields which exist as a patchwork of isolated elements and twisted flux tubes. If these emerge singly, they give rise to phenomena such as spicules, while if they burst through in concentrated bundles, they create the large-scale bipolar regions, sunspots and the multitude of varied forms of activity which keep the Sun in a state of turmoil.

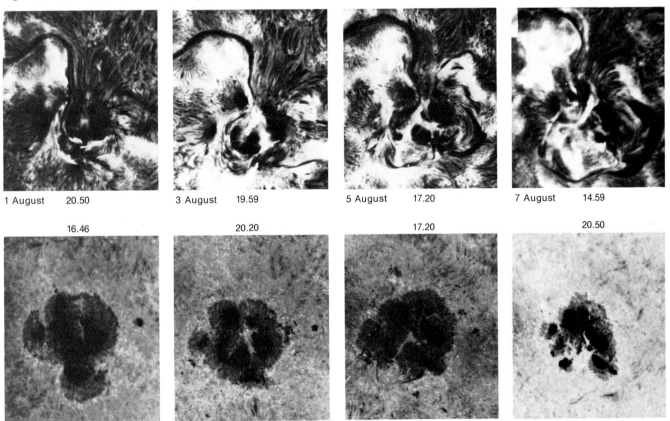

| 1 August | 20.50 | 3 August | 19.59 | 5 August | 17.20 | 7 August | 14.59 |
| 16.46 | | 20.20 | | 17.20 | | 20.50 | |

# Prominences and Filaments

Prominences are among the most beautiful and spectacular of solar phenomena. They appear at the limb of the Sun as flame-like clouds in the upper chromosphere and inner corona, and consist of clouds of material at lower temperatures but higher densities than their surroundings. Typical temperatures at their cores fall in the range of 10,000 K to 30,000 K (about one-hundredth of the coronal temperatures), while typical densities are about 100 times the value of the ambient corona. As a result, the gas pressure inside a prominence is roughly the same as that of its surroundings, although the mass of a given volume of prominence material will be about 100 times greater than that of an equivalent volume of coronal material at the same altitude.

Normally, prominences may be seen only in the light of certain Fraunhofer lines, such as $H_\alpha$, but during a total eclipse they may be seen directly in white light. In recent years space-borne instruments have made it possible for prominences to be studied in ultraviolet and even soft X-ray wavelengths. At X-ray and extreme ultraviolet wavelengths the prominences are often seen as dark shadows against the corona, since at these wavelengths the level of coronal emission is much higher than that of the relatively cool prominence material.

Filaments, which were formerly known as "dark flocculi", are the same type of phenomenon as prominences; the only difference between the two is the mode in which they are rendered visible, whether as emission or absorption features. Prominences are visible in emission at the limb because their relatively low temperature and high density ensure that compared with the background there is a concentration of non-ionized hydrogen atoms, which are responsible for $H_\alpha$ emission. At such wavelengths, therefore, the emission from the prominence far exceeds the background brightness of the corona. However, when seen against the bright background of the disc the absorption produced by these clouds predominates, and they are seen as dark filaments against the disc. (Any further comments made about either prominences or filaments may be taken to apply to both types of phenomenon.)

Prominences give the impression of being comprised of material surging upwards from the chromosphere, but in fact the great majority consist of material condensing from the corona and flowing downwards into the chromosphere, although prominences of the former kind do exist.

## Classification

Prominences may be divided into two principal types, quiescent and active, although these broad bands cover a wide variety of types and morphologies. Various classification schemes have been devised from time to time; the one shown on this page was devised in the 1950s by D. H. Menzel and J. W. Evans, with some modifications by F. Q. Orrall, and has the merit of giving a visual impression of the appearance of different classes of prominences.

## Quiescent prominences

Among the most stable and long-lived of solar phenomena, quiescent prominences may retain their overall shape and structure for periods ranging from a few months to as much as a year before breaking up, suddenly disappearing, or, occasionally, erupting violently into space and dispersing. Their typical dimensions are: length, about 200,000 km; height, 40,000 km; thickness 5,000 km to 8,000 km. They exist as long, thin, vertical sheets of material often taking on the appearance of a viaduct with numerous arches. There is considerable vertical structure and, although the overall shape remains remarkably constant, a considerable flow of material (usually downwards) takes place within the prominence structure. The "blade-like" appearance typical of many prominences is readily apparent in several of the photographs on this page.

Quiescent prominences occur along the neutral line separating regions of opposite polarity in a sunspot group, in active regions, or between larger-scale regions of opposite polarity. Along these lines the field lines run parallel to the solar surface, their slope changing from "up" to "down" as they flow from one side of the dipole to the other. Rather like elastic strings, they stretch under the weight of the dense prominence material, and the stretching can be regarded as increasing the tension, thus allowing the field to support the

**1. Prominence/filament**
As the Sun rotates, a prominence is carried over the limb and is observed as a filament on the disc. This sequence was recorded between 24 and 26 July 1980.

**2. Quiet filaments**
A crown of quiet filaments is often observed at times of minimum solar activity. This $H_\alpha$ photograph was taken on 4 October 1966.

1A 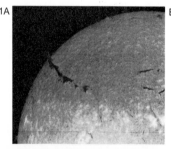 B 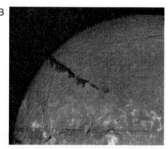 C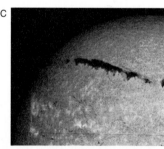

2.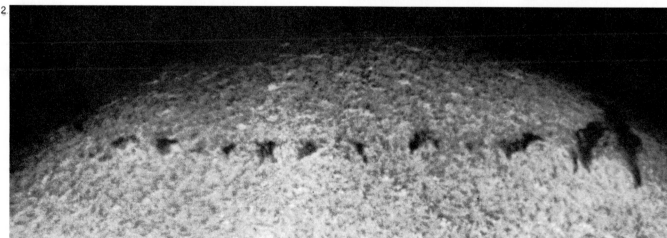

greater weight of the prominence, which is typically 100 times the weight of an equal volume of ambient coronal material. The horizontal field lines resist the vertical motion of ionized material. The magnetic field, then, supports the prominence and the structure, and any changes in the structure of the magnetic field are reflected in changes in the form of the prominence itself.

Closely related to quiescent prominences are the active region filaments, which form in the same kind of locations and also last for long periods; these, however, are characterized by the flow of material along their lengths, whereas quiescent prominences exhibit less movement of material, and what there is takes place mainly in a vertical direction. The filaments tend to form a few weeks after the formation of a spot group and then far outlive the spots themselves. Initially making an angle of about 40° with the local meridian, they become stretched out in length and make greater and greater angles with the meridian, approaching closer to an east–west direction with successive rotations because of the effect of differential rotation. On average a filament will increase in length by about 100,000 km per solar rotation, and they can stretch to lengths as great as 1,000,000 km.

After the spots have disappeared the filaments drift towards the poles. The frequency of occurrence of filaments and prominences roughly follows the sunspot cycle and, as far as equatorial and tropical ones are concerned, the latitudes at which they appear tend to follow Spörer's law (see page 40), forming progressively closer to the equator as the cycle develops. However, there is a phase difference of a couple of years between the maximum of sunspot activity and the maximum quiescent prominence activity.

Sudden changes can occur during the lifetime of a quiescent prominence. It may shrink and disappear or suddenly erupt outward. Often, after the disappearance, a prominence will re-form in a closely similar location. Sometimes when a prominence disappears in chromospheric temperature and light (such as $H_\alpha$ at, say, 10,000 K) it may then appear at a higher altitude and at a much higher temperature in the extreme ultraviolet, as if the material had ascended and been rapidly heated.

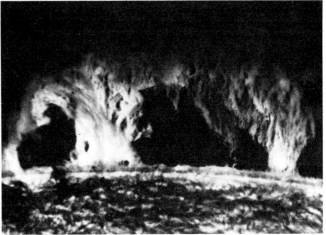

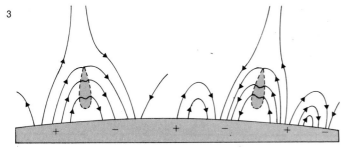

**3. Prominence support**
The relatively dense material of a prominence appears to be supported by the local magnetic field. Two possible field configurations are shown in the diagram. The support results from a force that is produced by a current flowing perpendicular to the direction of the magnetic field lines. This current in turn produces a magnetic field that distorts the shape of the original field lines.

**Classification of prominences**

A. Prominences originating from the corona
    S. Spot prominences
        a. Rain
        f. Funnels
        l. Loops
    N. Nonspot prominences
        a. Coronal rain
        b. Tree trunk
        c. Tree
        d. Hedgerow
        f. Suspended mound
        m. Mound
B. Prominences originating in the chromosphere
    S. Spot prominences
        s. Surges
        p. Puffs
    N. Nonspot prominences
        s. Spicules

**4. Hedgerow prominence**
The vertical structure of this small limb prominence, photographed in $H_\alpha$ on 7 December 1970, remains static while material from the corona flows downwards through filamentary "tubes".

**5. Arch prominences**
This series of arches seen in $H_\alpha$ light rises to a height of over 65,000 km above the level of the photosphere.

**6. Loop prominence**
The loop structure of this prominence outlines the magnetic field above a sunspot group. In spite of its appearance, which suggests material flowing upwards, this prominence, like those in the previous two illustrations, in fact consists of material condensing out of the corona and flowing downwards. It is seen here in the light of the green line due to FeXIV.

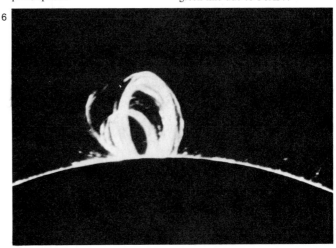

# Active Prominences

Shorter lived and usually smaller than quiescent prominences, active prominences may show dramatic changes in form within minutes. They have a mean length of about 60,000 km. They are normally associated with sunspot groups and have two basic characteristic forms: arches (or loops) on the one hand, and condensation and knots on the other. In loop prominences and "coronal rain", material from the corona descends, some of this material having previously been ejected from the chromosphere. In developing active regions, arch prominences appear at a fairly early stage as a result of the emergence of magnetic flux from below the surface. In arch prominences of this type, material descends also from above, flowing down through both feet of the arch. Arch prominences connect regions of opposite polarity in developing sunspot groups, running across the neutral line, unlike quiescent prominences which run along it. The lifetime and content of these short-lived features are controlled by the rate of condensation of material from the corona and the rate of flow along the magnetic field lines into the chromosphere. Depletion of the source of material and changes in the local magnetic field can terminate its existence.

In contrast, surges, puffs and sprays consist of material being hurled upwards from below. These are the most violently active of prominence phenomena and are mostly related to flares: in surges which occur in active regions, chromospheric material is shot into the corona at speeds of around 100 to 200 km s⁻¹. Thousands of

millions of tonnes of matter can soar to heights of several hundred thousand kilometers before falling back, usually along the same track as the ascent. The falling material may break up and be lost from sight before it reaches the surface. The typical lifetimes of surges are about 10 to 20 min, and they frequently recur. Puffs often occur just prior to surges, and take the form of brief sudden expansions of material in flare regions.

Sprays are even more violent impulsive events, in which chromospheric material is ejected in excess of the solar escape velocity (618 km s⁻¹) and is scattered widely. They are essentially more energetic versions of surges, and are invariably associated with the occurrence of flares.

Magnetohydrodynamic waves (*see* page 35) emitted from flares may also set prominences into vertical oscillation and may trigger their disappearance, although in many cases another prominence of similar shape will re-form in the same region. Prominences may erupt, following a flare outburst, usually as a growing arch whose center expands rapidly and disappears while the ends remain visible and rooted in the chromosphere below.

Extreme ultraviolet studies made with Skylab instrumentation have indicated that loop prominences consist of a cooler inner core (at around 20,000 K) surrounded by a hotter sheath at 100,000 K or more revealed in ultraviolet emission lines. However, there remains doubt as to whether or not this picture applies to all prominences, and further analysis and observation are required.

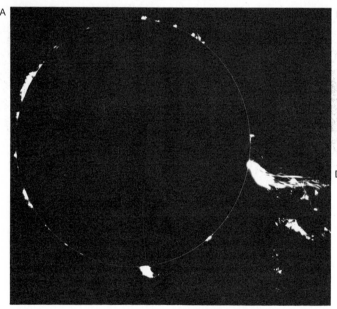

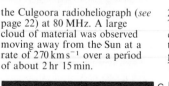

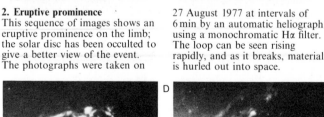

**1. "Westward Ho"**
This eruption, nicknamed the "Westward Ho" event, occurred on 2 March 1969. The ejection of solar gas is shown in Hα (**A**) together with a sequence made by

the Culgoora radioheliograph (*see* page 22) at 80 MHz. A large cloud of material was observed moving away from the Sun at a rate of 270 km s⁻¹ over a period of about 2 hr 15 min.

**2. Eruptive prominence**
This sequence of images shows an eruptive prominence on the limb; the solar disc has been occulted to give a better view of the event. The photographs were taken on

27 August 1977 at intervals of 6 min by an automatic heliograph using a monochromatic Hα filter. The loop can be seen rising rapidly, and as it breaks, material is hurled out into space.

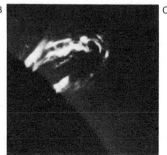

# Color Plates

Solar corona
Voi, Kenya, East Africa
16 February 1980

Visible solar spectrum
Sacramento Peak Observatory

Flash spectrum
Pacific Ocean
12 October 1977

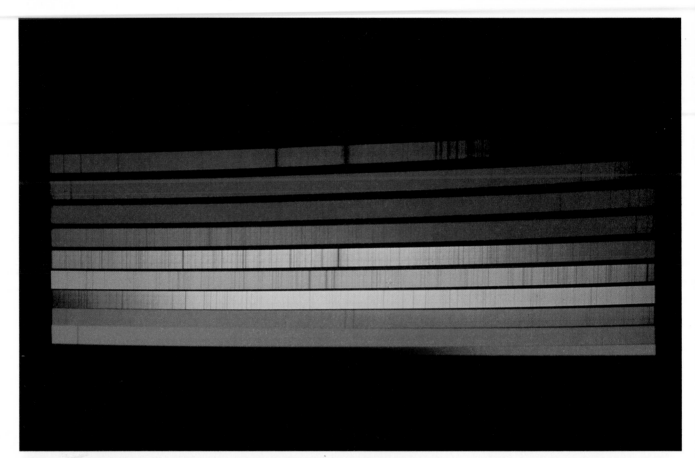

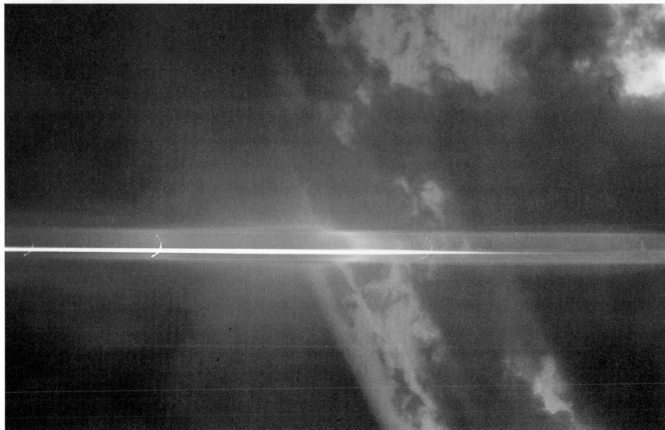

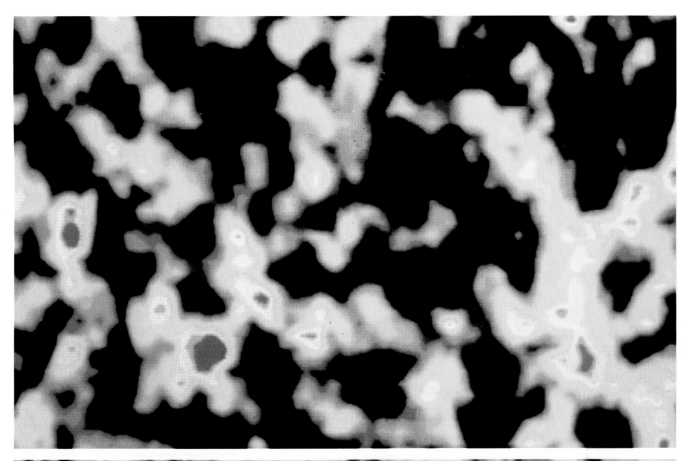

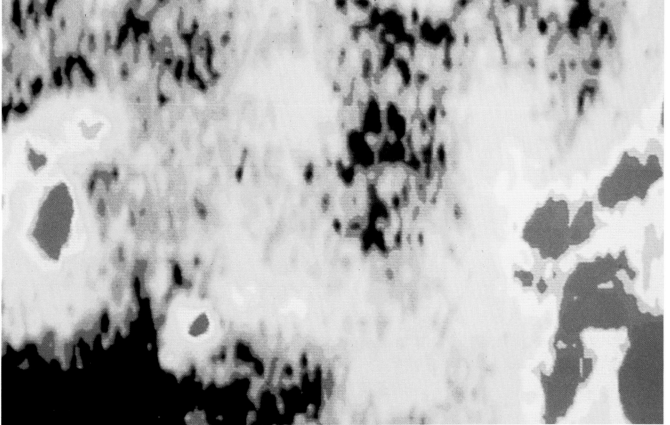

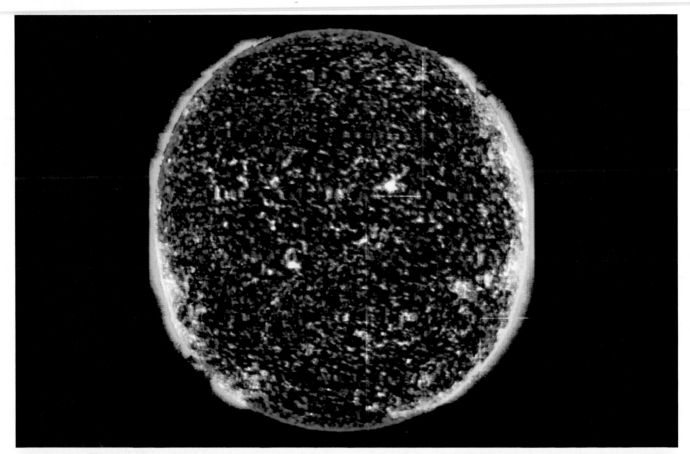

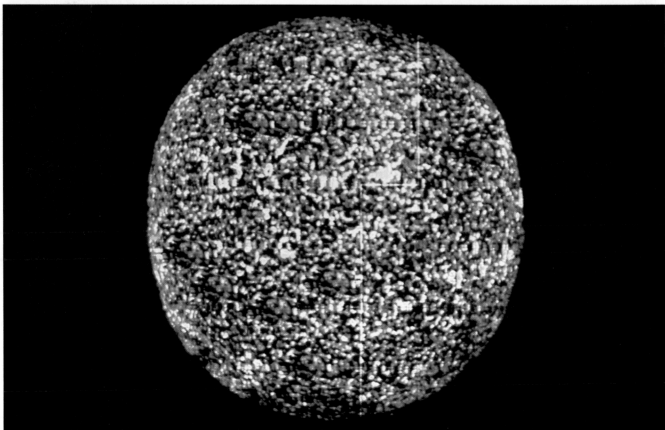

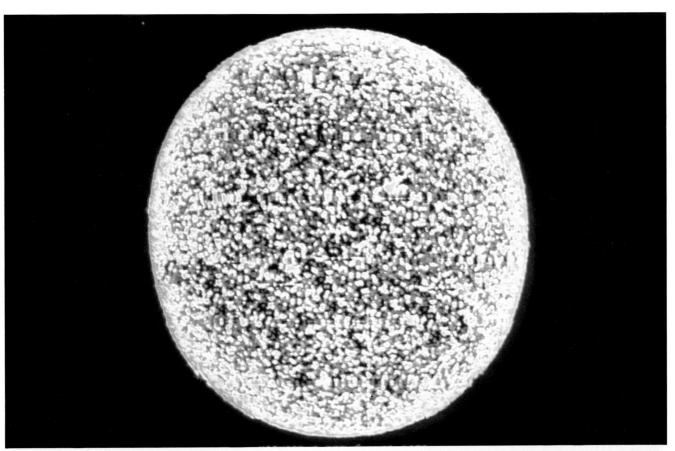

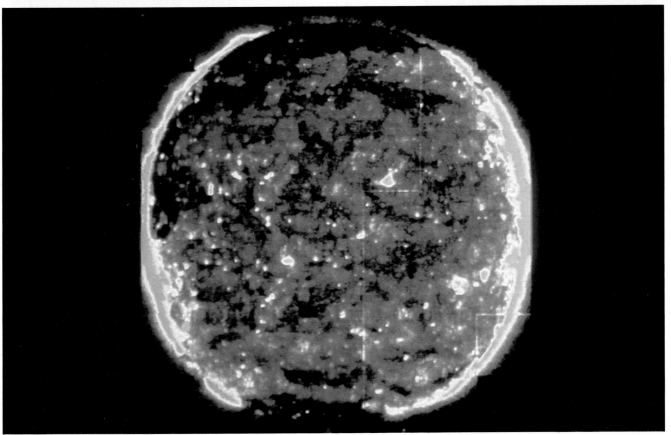

**Loops of active region**
**Skylab**
**4 December 1973**

**Magnetic loop**
**Solar Maximum Mission**
**27 March 1980**

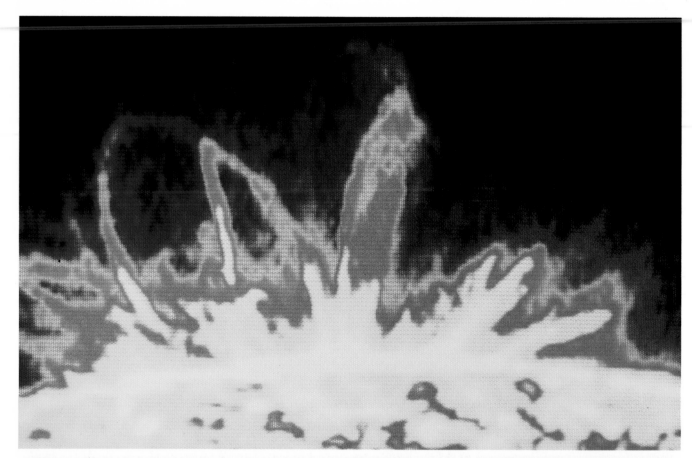

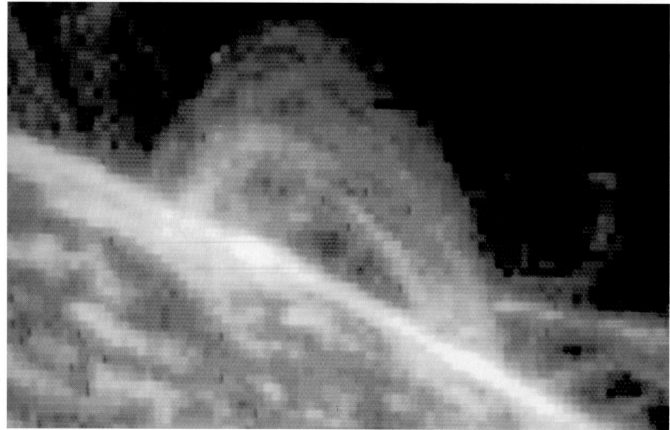

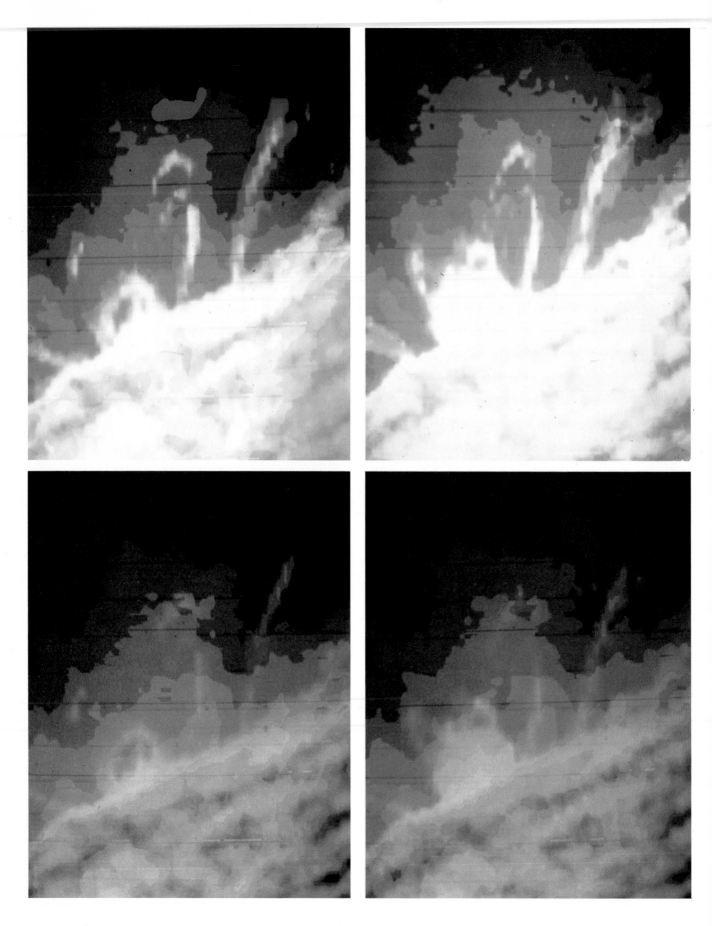

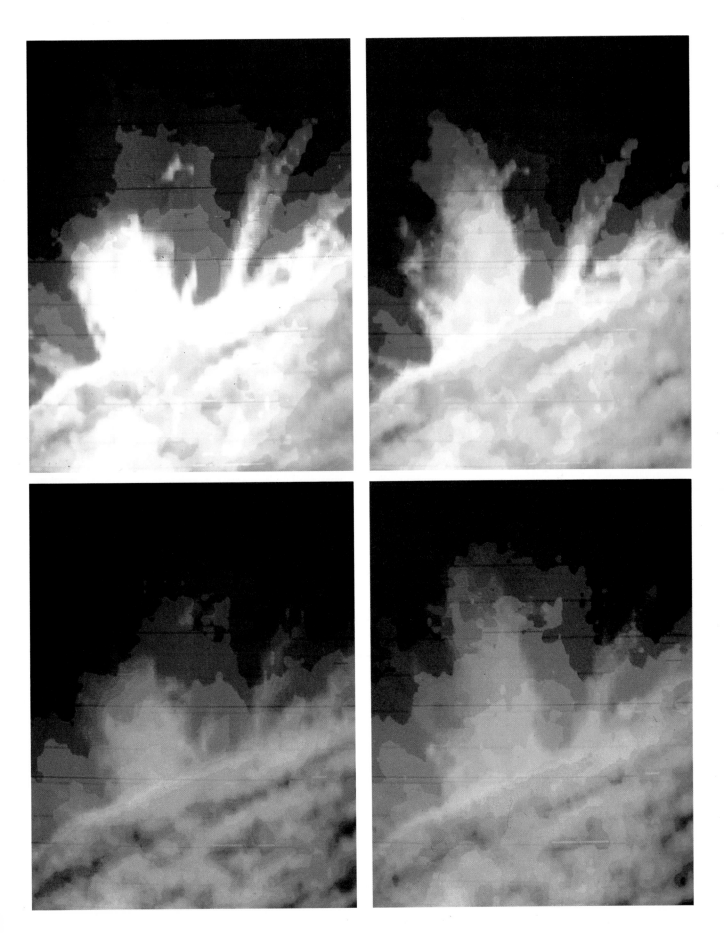

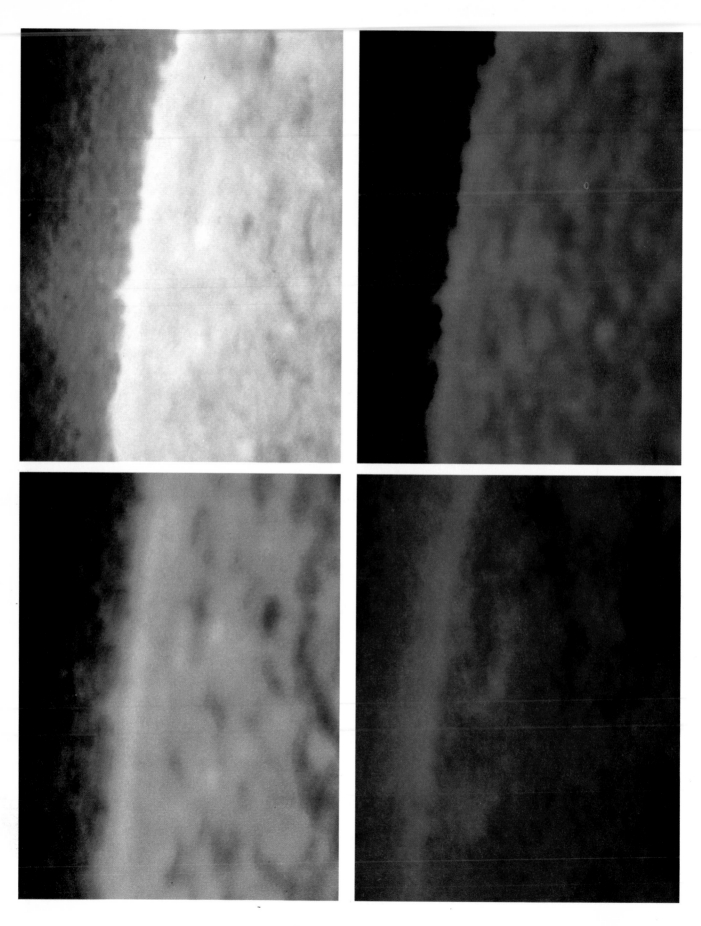

Active region
Skylab
14 August 1973

X-ray Sun
Skylab
16 June 1973

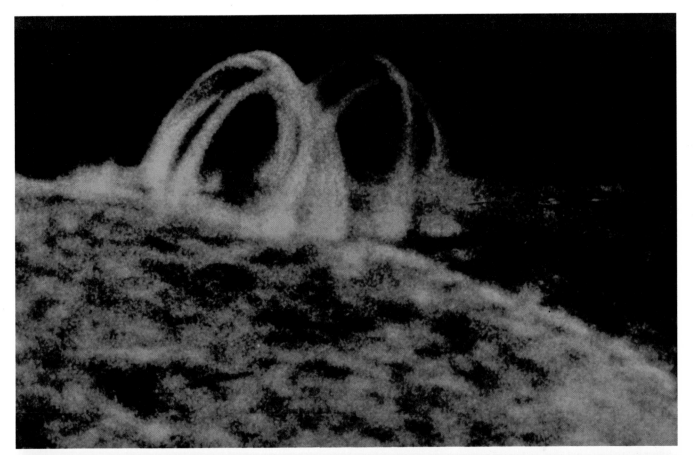

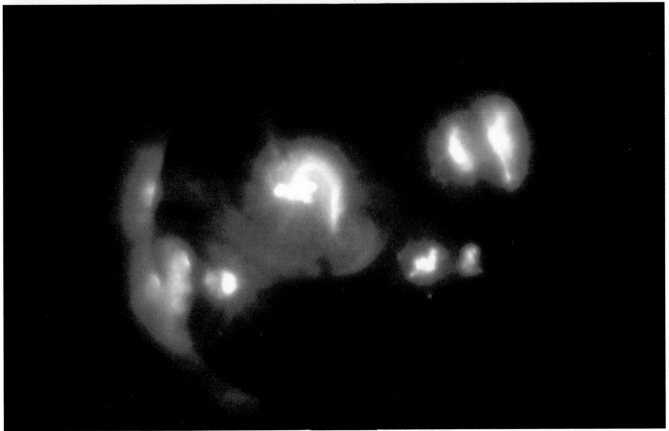

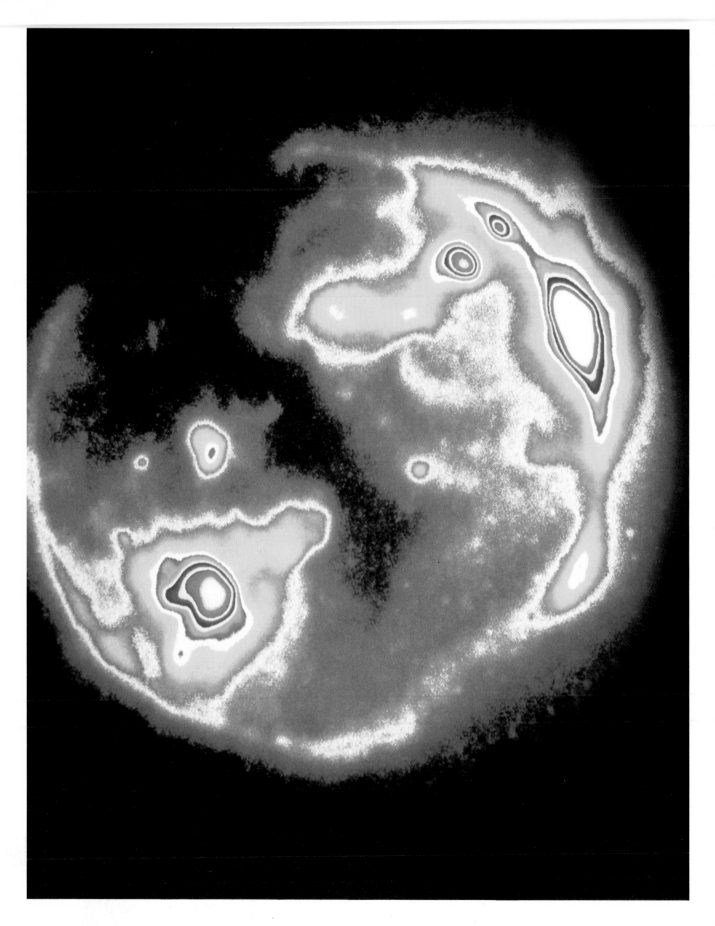

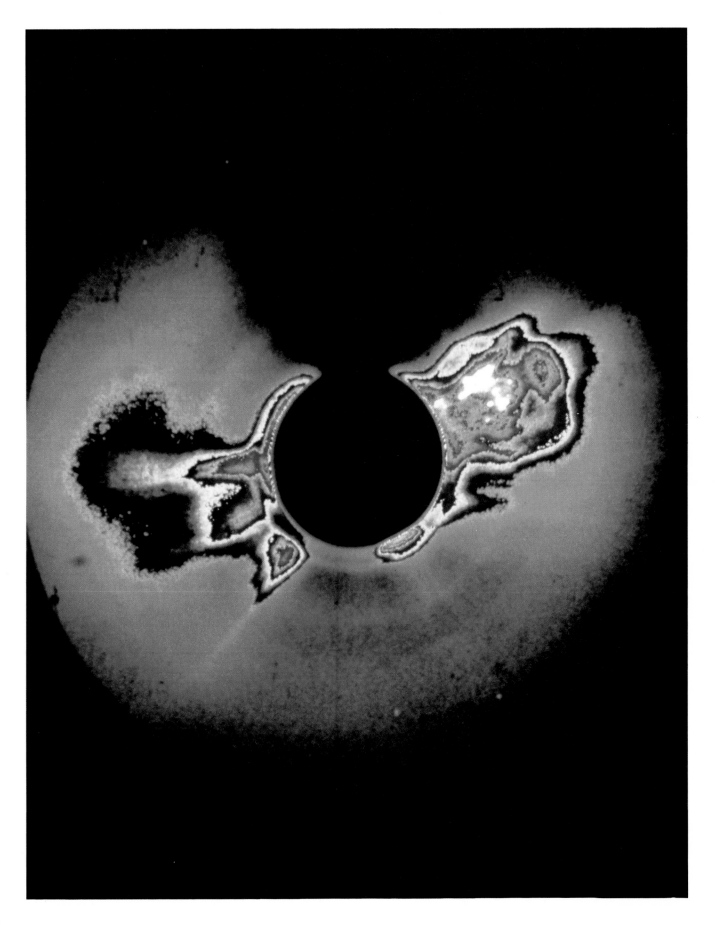

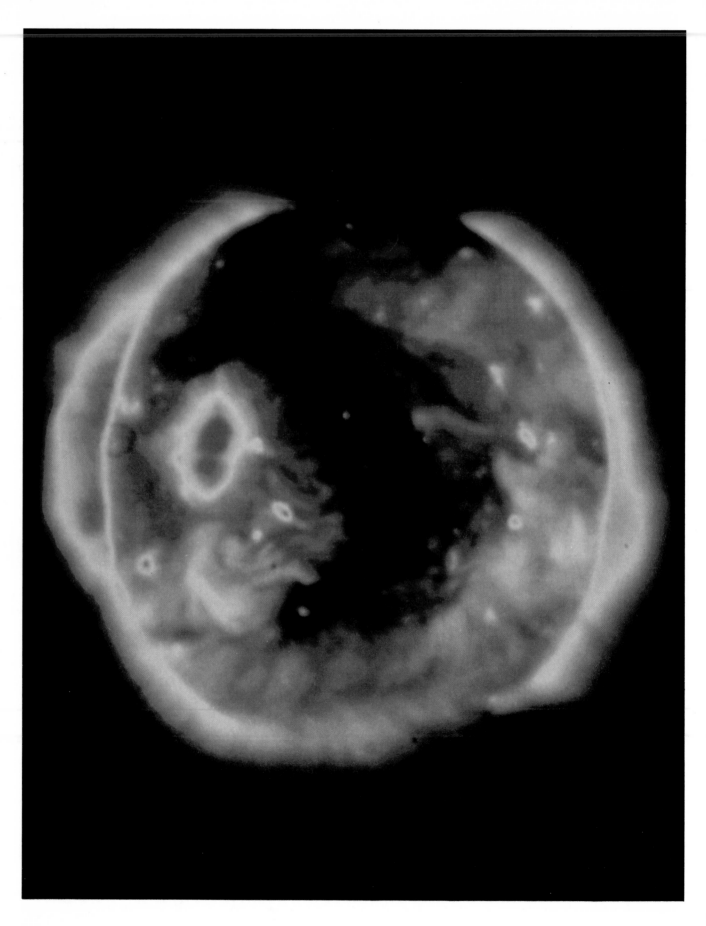

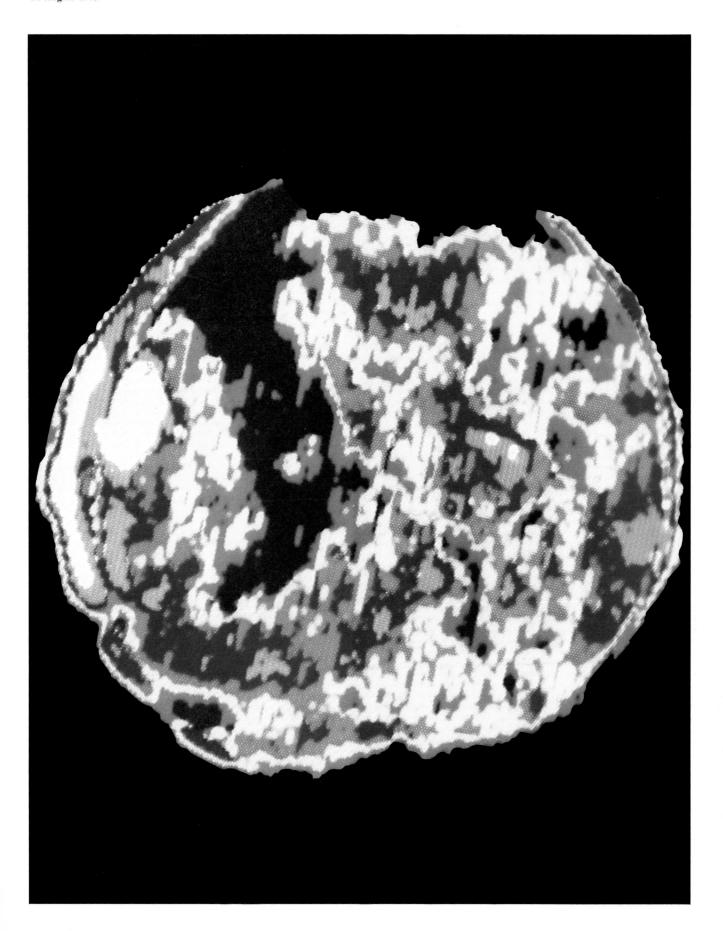

X-ray image
Skylab
30 June 1973

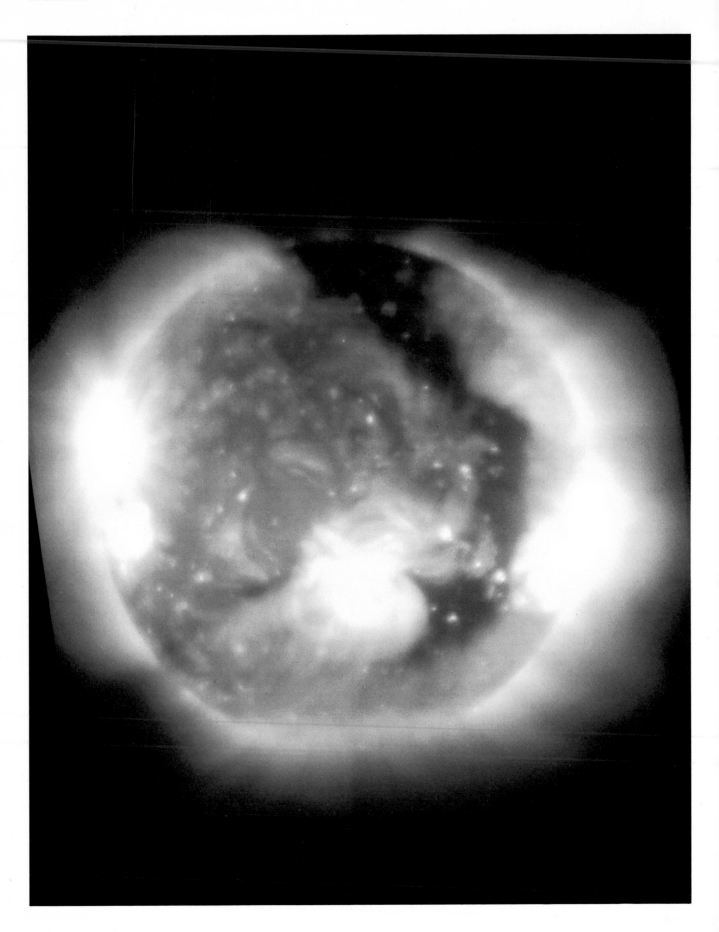

## Notes to color plates

**Page 49.** Much fine structure is visible in these magnificent photographs of the corona. Several prominences can also be seen.

**Page 50.** (Top) The visible portion of the Sun's spectrum reveals hundreds of dark absorption lines, each produced by a particular element at a particular level of ionization. For example, the two very broad lines near the center of the top row (in the violet region) are characteristic of singly ionized calcium (Fraunhofer's H and K lines). (Bottom) A number of chromospheric emission lines can be seen in this spectrum photographed at second contact: $H_\gamma$ in the dark violet (434 nm), $H_\beta$ in the blue (486.1 nm), the strong $D_3$ line due to neutral helium in the red, and $H\alpha$ (656.3 nm). The bright horizontal line is the result of light from a Baily's bead.

**Page 51.** These two ultraviolet images illustrate the way in which the chromospheric network becomes indistinct at high altitudes (*see* page 33). At a wavelength of 97.7 nm (top), representing chromospheric temperatures of around 20,000 K, the network is clearly defined; in the coronal light of 62.5 nm (bottom)—representing a temperature of about $1.4 \times 10^6$ K—the pattern has virtually disappeared.

**Page 52 and 53.** The first image (top left) is a composite of the other three, each of which was taken at a different ultraviolet wavelength: 121.6 nm (bottom left), 97.7 nm (top right), and 62.5 nm (bottom right). 121.6 nm is the Lyman-alpha line of hydrogen; this image shows the chromospheric network in the middle chromosphere. The next image (a wavelength due to CIII) represents a level higher in the chromosphere; the spicules are seen end-on, and appear as bright speckles. The final image is in the light of MgX, and shows the corona, with coronal holes at the polar regions (*see* pages 74–75).

**Page 54** (Top). This false-color image shows ionized material following the magnetic loop patterns above an active region. The wavelength is 103.2 nm, due to the light of OVI. (Bottom). A similar loop appears in this SMM image taken at a wavelength of 154.8 nm (CIV).

**Page 55.** An eruptive prominence (*see* pages 46–48) is shown here in a computer-enhanced image taken in the light of the HeII (30.4 nm).

**Pages 56 and 57.** Two sequences (reading from left to right across the page) show the rapid development and eruption of chromospheric material breaking loose of the Sun's magnetic loops. Both sequences are computer-enhanced ultraviolet composites.

**Page 58.** An ultraviolet composite (top left) with three component images: 97.7 nm (CIII), 103.2 nm (OVI), and 62.5 nm (MgX). The high chromosphere (top right) has been colored blue, and shows sharply defined spicules at the limb. The transition region between the chromosphere and corona has been colored green (bottom left), while the corona itself (bottom right) is shown in red.

**Page 59** (Top). A weaker secondary image appears next to the predominant NeVII image at a wavelength of 46.5 nm. (Bottom). High-temperature X-ray emissions from the corona, reaching a maximum of over $5 \times 10^6$ K, appear concentrated along a band near the equator. The bright curve near the center represents a huge arch.

**Page 60.** In this X-ray image, false colors have been used to represent different temperatures, with white showing the highest temperature. A large coronal hole (black) extends downward from the north polar region.

**Page 61.** The temperature structure of the corona is revealed in this false-color image of a white light coronagram.

**Page 62.** An X-ray image of the solar disc shows a large "S-shaped" coronal hole extending from the Sun's north polar region towards the equator.

**Page 63.** A false-color enhancement of an extreme ultraviolet image showing the same coronal feature as that seen in the X-ray image on page 62.

**Page 64.** This X-ray image shows the corona as it appeared a month before the images on pages 62 and 63. The developing coronal hole is already apparent.

# Flares I

Flares are sudden violent releases of energy that occur in the vicinity of active regions. They eject atomic particles and emit radiation across the entire range of the electromagnetic spectrum from γ-rays and "hard" X-rays (wavelengths less than $10^{-10}$ m) to radio wavelengths of several kilometers. The first flare to be recorded was seen by the English astronomer Richard Carrington, in 1859; it was visible in white light, but it is only relatively rarely that a "white light flare" can be seen against the brilliance of the photosphere. Flares are normally seen by Earth-based observers in monochromatic light, particularly $H_\alpha$ and K-calcium.

A flare consists of an $H_\alpha$ region emitting at temperatures of the order of 10,000 K, embedded within which are smaller bright regions or "kernels" some 1,000 to 10,000 km in diameter. High-temperature flare emission is seen in extreme ultraviolet wavelengths (1 to 100 nm). The individual sources of this radiation usually take the form of small arch-shaped regions, typically less than 10,000 km in size.

Flares usually occur close to the neutral line in complex active regions. In $H_\alpha$ a flare is generally seen to commence in a number of compact bright points of light within a plage, and the region of enhanced emission rapidly spreads—within a few minutes—to cover an area typically of some $10^9$ km². Where two or more bright spots are seen, they are found to lie on opposite sides of the neutral line and, in larger flares, the individual bright patches merge into two rows giving rise to a "two ribbon" flare. After the disturbance caused by the flare, the magnetic field sometimes appears to restore itself approximately to its original configuration, since after a few hours a similar flaring phenomenon may recur in the same site.

The $H_\alpha$ emission results from a temperature increase of several thousand Kelvin in a thin region at upper chromospheric or low coronal altitudes; observations made at the solar limb indicate that flares typically reach heights of 5,000 to 15,000 km above the photosphere. The flare reaches maximum brightness within a few minutes and declines more slowly thereafter, the total duration of such events ranging from about 10 min to several hours, with 20 min being typical for smaller flares. In $H_\alpha$, flares are classified in order of importance according to the area they cover and the maximum brilliance attained (see Table 1). Events smaller than 2 square degrees—angles measured about the Sun's center—on the solar surface (less than $3 \times 10^8$ km² or 100 millionths of the area of the visible hemisphere) are known as subflares. By an alternative system, flares are classified according to their peak brilliance in soft X-rays (0.1 to 0.8 nm), giving a better guide to the energy output and resulting terrestrial effects (see Table 2).

The numbers of flares occurring is closely related to the solar cycle. An approximate guide to the daily frequency of flares of importance class 1 or greater is given by R ÷ 25, where R is the Wolf Relative Sunspot Number (see pages 40–41). For example, at a high maximum with R equal to 150, there would on average be $150/25 = 6$ flares per day, while at minimum several days could pass without

1A

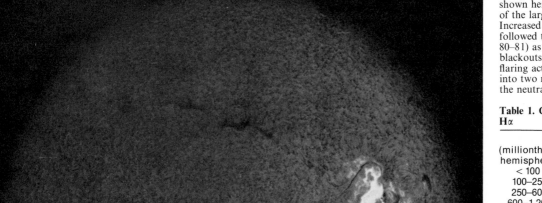

### 1. Flare
The great flare of 7 August 1972, shown here at its peak, was one of the largest ever recorded. Increased auroral activity followed the flare (see pages 80–81) as well as shortwave radio blackouts. The major part of the flaring activity is concentrated into two ribbons on either side of the neutral line.

### Table 1. Classification of flares – Hα

| Area (millionths of hemisphere) | (square °) | Class |
|---|---|---|
| < 100 | < 2.0 | s |
| 100–250 | 2.0–5.1 | 1 |
| 250–600 | 5.2–12.4 | 2 |
| 600–1,200 | 12.5–24.7 | 3 |
| > 1,200 | > 24.7 | 4 |

The maximum brightness of a flare in Hα is added to the importance class: F = faint, N = normal, and B = bright. The least important flare is 1F and the most important 4B.

### Table 2. Classification of flares – soft X-rays

Peak flux, $\phi$, in the range 0.1 to 0.8 nm

| | Class |
|---|---|
| $\phi < 10^{-5}$ W m$^{-2}$ | C |
| $10^{-5} \leqslant \phi < 10^{-4}$ | M |
| $\phi > 10^{-4}$ | X |

A one-digit number from 1 to 9 is added as a multiplier. Thus a C5 event has a flux of $5 \times 10^{-6}$ W m$^{-2}$ and X7 corresponds to $7 \times 10^{-4}$ W m$^{-2}$. The class C0 is used for events below $10^{-6}$ W m$^{-2}$. This classification dates from 1 January 1969.

any flare activity. The average frequency of white light flares is less than one per year.

Prior to the onset of a flare, adjustments take place in the magnetic field of the region in which it is about to occur. For example, if a quiescent filament is present close to the location it will be triggered into activity. The first phase of some flares, and certainly of the more energetic ones, is a sharp burst of hard X-rays followed by a more gradual rise and fall of soft X-rays. Simultaneous flashes may be seen in optical and extreme ultraviolet wavelengths; SMM results seem to show that the hard X-ray bursts coincide in time with the sudden intensification of ultraviolet emission near the tops of small loop structures within the flare region while the hard X-rays themselves come from the feet of the loop. The temperature in flare kernels may be as high as $10^9$ K.

The hard X-rays are produced by streams of electrons that are rapidly accelerated to speeds of up to half the velocity of light; the radiation is produced as a result of close collisional encounters between these "relativistic" particles and the surrounding particles in the solar atmosphere. Radiation of this type, emitted by electrons that are sharply slowed down, is known as "bremsstrahlung" radiation. Strongly polarized microwave radiation is also emitted by the "synchrotron" process (see page 29) as the streams of electrons move out from the flare site. The microwave bursts span a wide range of frequencies from above $10^{10}$ Hz down to about $10^8$ Hz (wavelengths from below 0.01 m to about 1 m).

**2. Development of a flare**
These 12 filtergrams, taken on 8 August 1968, show the development of a Class 2B flare. The center of the Hα line (**A**) gives the clearest view of the flare. Off-center wavelengths (**B** and **C**) allow the sunspot group at the center of the flare region to be observed in greater detail.

**3. Ultraviolet flare spectrum**
Part of the ultraviolet spectrum of a solar flare was photographed by Skylab on 15 June 1973. Thousands of emission lines (black in this negative print) appear between 115 and 175 nm. The broad band centered on a wavelength of 121.6 nm is the Lyman-α line of hydrogen.

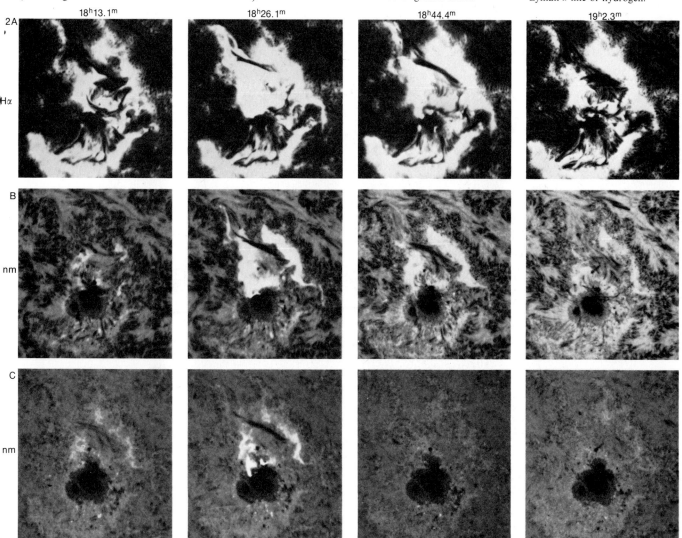

67

# Flares II

High-speed streams of electrons and sometimes protons may be ejected from flares, together with bulk masses of material. The particles travel outward through interplanetary space and have important effects on the terrestrial magnetosphere and atmosphere (*see* pages 78–81). Streams of low-speed (non-relativistic) electrons are also emitted; a typical large flare event expels up to $10^{39}$ of these electrons. The high-speed electrons are emitted in smaller numbers (say $10^{36}$ in a large flare) and a similar number of protons is ejected in a major flare.

As the fast electrons accelerated by the flare event pass out through progressively more rarefied layers of the solar atmosphere, they set the coronal plasma oscillating in such a way as to produce radio emissions. Solar radio emissions at wavelengths of meters or tens of meters are classified into five major types of "noise" bursts.

Type I noise storms—which are not specifically associated with flares—consist of long series of bursts, each individual event appearing as a spike of 0.1 to 2 sec duration superimposed on a general increase in solar radiation at these frequencies. These noise storms may last from a few hours up to several days. Type II bursts, in the 300 to 10 MHz range to 1 to 30 m wavelength, usually occur 10 to 30 min after the onset of a flare and are due to the effects of a shockwave ascending through the corona at speeds in the region of 800 to 2,000 km s$^{-1}$. Type III bursts, in the 800 MHz to 10 MHz range occur typically 2 to 4 min after the onset of a flare. They consist of short-lived spikes that drift in a few seconds from high to low frequencies, and are therefore referred to as "fast drift bursts". The source of these bursts are plasma oscillations generated by the high-speed electrons moving at up to half the speed of light.

Type IV bursts are complex in form, spanning a wide range of frequencies. The radiation arises by the synchrotron process, and events of this type usually occur after the Type II bursts in large

flares and proton flares (flares that emit energetic protons and other heavy particles—solar cosmic rays). Part of the burst moves out through the corona with speeds in the range 100 to 1,400 km s$^{-1}$; these "moving type IV bursts" are emitted from a blob of hot plasma moving outward and carrying with it its own magnetic field. Such bursts have been shown to follow in the wake of loop-shaped coronal transients (*see* pages 74–75). Type V bursts are similar to Type III, occurring a little later, and might be caused by the trapping or delaying of some of the high-speed electrons.

The sequence of events in a flare would seem to be as follows. Over a period of hours or days prior to the event large amounts of energy are built up and stored in the magnetic field over a complex active region. Somehow, after activation of some trigger mechanism, the energy is released very rapidly indeed and as a result streams of electrons are accelerated to near-relativistic speeds; the interactions of these electrons with the magnetic field and the surrounding plasma give rise to a wide variety of emissions from X-rays to radio waves. The sudden heating of local areas of the chromosphere and corona results in the expulsion of considerable masses of material, up to several thousand million tonnes, which may move out into interplanetary space. Shock waves spread out through the photosphere, chromosphere and corona at speeds of 1,000 to 2,000 km s$^{-1}$, setting up violent oscillations in prominences up to hundreds of thousands of kilometers distant.

## Theories of flares

Despite decades of intensive research there is still no clear accepted theory that accounts satisfactorily for all, or even any, of the aspects of the flare mechanism. There is no doubt, however, that the primary energy source is stored magnetic energy. No other source can provide the necessary energy in a concentrated volume over the

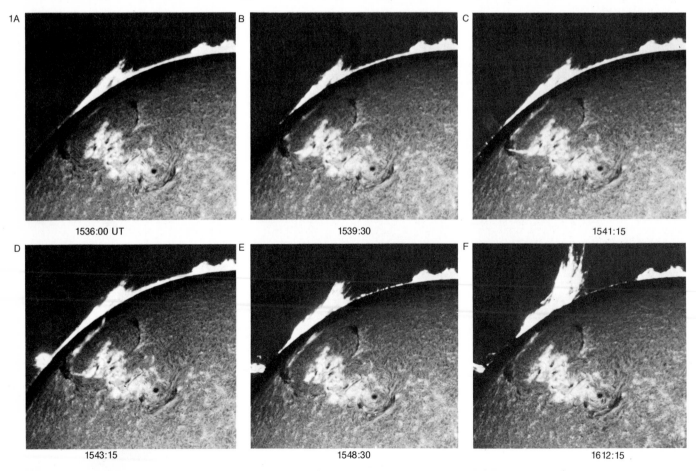

1A    1536:00 UT

B    1539:30

C    1541:15

D    1543:15

E    1548:30

F    16 12:15

short timescale involved. The rate at which energy is released from a typical flare is about $10^{21} \text{J s}^{-1}$ (about $10^{-5}$ of the solar luminosity). Exceptionally, flares can emit at up to $10^{-3}$ of the entire solar output. If the duration of the flare is assumed to be about 1,000 sec, then its total energy output would be some $10^{24} \text{J}$. Certainly, there is plenty of magnetic energy available; it has been shown that the $10^{25} \text{J}$ required for a large flare could be attained by depleting the field strength of a 3,000 G sunspot group by about 20 percent over a volume of $10^{12} \text{km}^3$.

Flares occur much more frequently in magnetically complex active regions than in simple bipolar groups, and there is no doubt that the complexity of the magnetic field and the presence of warped, twisted, neutral lines is an important factor in flare production. Studies have shown that the number of flares seen in $H_\alpha$ in an active region is closely correlated with the number of kinks in the neutral line. The emergence of new flux from below the photosphere in an existing active region further enhances the complexity of the region and induces flare activity. The largest single X-ray emission spike so far detected (by the SMM on 29 March 1980) was inferred to originate from the site of emerging magnetic flux of opposite polarity to the existing field.

There have been many theories as to the form of magnetic field that leads to the sudden energy release, but it is generally agreed that field "reconnection" or "annihilation" is the process best able to release the requisite quantities of energy (*see* diagram 2).

According to one model originally proposed by P.A. Sweet and by E.N. Parker, above a bipolar region and between the regions of opposite polarity there can exist a "neutral sheet" separating lines of force with opposite directions. A slight instability is sufficient to bring together such lines so that they join up, or reconnect, with the sudden release of energy and the expulsion of particles and plasma.

According to an alternative point of view, advocated primarily by D.S. Spicer, a twisted arch (or loop) of magnetic flux joining regions of opposite polarity provides the source and site of the energy release. Instabilities in the tube lead to the breaking up of the tube into many magnetic "islands", and the resultant reconnection of field lines converts magnetic energy into the requisite form for flare phenomena. This process is known as the "tearing mode instability". However, in neither picture is it as yet possible to release energy fast enough to account for observations.

The magnetic arch picture tends to be favored by extreme ultraviolet and X-ray observations, which show that flares occur either in single arches or in a series of arches (an "arcade"). SMM observations indicate that these arches are the sites of X-ray, ultraviolet and $H_\alpha$ bright points at the onset of flares.

Even with an appreciation of the basic mechanisms of energy release there are still formidable difficulties in the way of a full understanding of such problems as how the energy is channelled into the appropriate forms and how the particle acceleration takes place. The whole subject is immensely complicated, and the difficulties experienced in trying to account for such phenomena on the Sun indicates how important it is to exercise great caution when making statements about the physics of distant stars.

The study of flares is of the utmost interest for a wide variety of reasons: the X-ray, ultraviolet and particle emissions have direct effects on the Earth, and large flares pose hazards for astronauts in space; an understanding of the flare mechanism would provide vital clues in the search for energy from controlled thermonuclear fusion by magnetic confinement on Earth, with important implications for the global energy crisis. The same kind of process is thought to be responsible for the behavior of flare stars—faint red stars which, from time to time, flare up to several times their normal brightness.

**1. Flare sequence**
Two superimposed images show the development of a small flare with an associated spray on 21 May 1967.

**2. Flare mechanism**
According to one model (**A**) a neutral sheet exists between magnetic field lines with opposite directions above a bipolar magnetic region. Reconnection of field lines results in a sudden release of magnetic energy (**B**) with the expulsion of streams of particles. A large blob of plasma may ascend through the corona (**C**) carrying magnetic flux with it. An alternative possibility (**D**) for the site of the primary release of flare energy is the neutral point between an established loop of magnetic flux (the larger loop) and a newly emerging loop. A third possibility (**E**) is that flaring occurs in magnetic arches connecting regions of opposite magnetic polarity. Theory indicates that the field lines in such an arch will follow helical paths and that flaring can occur at many points along the arch.

**3. Radio noise storms**
The frequencies of the various emissions have been plotted against time elapsed since the onset of a flare. The diagram is schematic, and not all flares are necessarily followed by all types of noise storm.

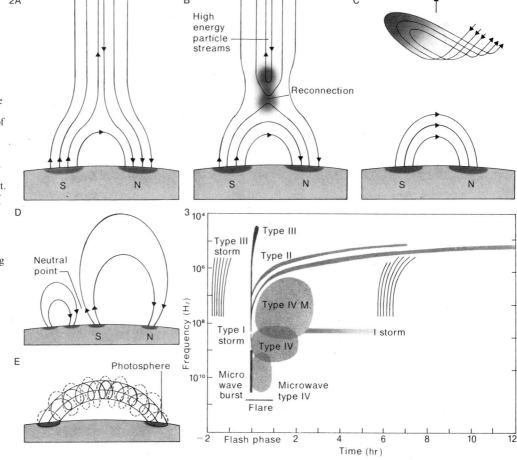

# The Corona I

The corona is the outer atmosphere of the Sun. It emits on average slightly less visible light than the full Moon but, because of the level of scattered sunlight close to the Sun in the sky, it cannot be seen with the unaided eye except during a total solar eclipse. In the 1930s the invention of the coronagraph (*see* pages 20–21) allowed the inner corona to be studied at other times from high-altitude sites when conditions were favorable, but it is only since the early 1970s, when it has been possible to place coronagraphs in orbit, that the corona could be studied continuously under ideal conditions. Another major development has been the growth of extreme ultraviolet and X-ray techniques, which have allowed the corona to be studied all over the solar disc rather than merely at the limb since the background (cool) photosphere does not swamp the (hot) corona at these wavelengths.

The appearance of the corona can vary markedly, since its overall shape changes with the solar cycle. At sunspot minimum the corona typically has a fairly symmetrical structure with long streamers extending outward from the equatorial regions, while polar plumes emerge from the polar regions rather like the pattern adopted by iron filings in proximity with the poles of a bar magnet. At maximum the corona is more conspicuous and usually rather more evenly spread over the whole disc.

Long extensions of the corona ("streamers") are found over active regions of the photosphere, and may take the form of "fans" (broad extensions usually found over quiescent prominences) or of narrower "rays". Both types of feature clearly follow the geometry of the local magnetic fields. In the low corona, arch-shaped coronal condensations are found, following closed arches or loops of magnetic field between regions of opposite polarity within active regions, or larger arches spanning regions of opposite polarity in separated regions. High above such arches extensive streamers may be found following the open pattern of field lines out into interplanetary space. The frozen-in plasma can readily move along lines of force (*see* pages 36–37), but motion across the field lines is actively resisted. The density of material in these concentrations is typically five to ten times that of the ambient corona. Coronal arch systems that are not associated with streamers are also found. Around prominences and loops the corona appears darker, due to the decrease of density that occurs as material is concentrated by the field into the condensations.

The general distribution of fans, rays and other coronal features follows the latitude of solar active regions. Thus, at solar minimum the streamers tend to be concentrated in equatorial regions, while at maximum they are more widely spread in latitude. The greatest extensions occur above sunspots, where the extensions at minimum have been observed out to colossal distances, over 20 $R_S$ on occasion.

## X-ray and extreme ultraviolet observations
The detailed structure of the corona and the degree of influence exerted by magnetic fields has only become apparent since the era of rockets and spacecraft-borne instrumentation, which have made it possible to look deep into the ultraviolet and X-ray part of the spectrum. Because of its immense temperature, the hot coronal gas is a strong emitter of X-rays, and by examining these wavelengths, concentrations and structures in the corona are rendered visible over the entire solar disc. When observed in this way the corona is

1

seen to consist of three principal types of regions:
1. Active regions and coronal bright points, which are regions of bright X-ray emission associated with strong, closed loops of magnetic field.
2. Quiet regions, of much fainter X-ray emission, with weak coronal magnetic fields which nevertheless appear to be closed on a large scale; in general they are larger and fainter than active regions.
3. Coronal holes (*see* pages 74–75).

### Magnetic influence
Magnetic loop structures seem to be the basic building blocks of the entire corona. Material is concentrated to higher densities and temperatures within these structures, while open lines spread out over the tops of the closed regions and then stream out into space. The underlying loop, overlaid by the open field lines, gives coronal streamers their characteristic bulb shape above active regions.

The basic magnetic connections between active regions, shown up as loops at X-ray wavelengths, seem to persist for the duration of the lifetime of those active regions, but the individual loops survive for much shorter times, usually less than a day. It has been said that the corona can be considered to be a collection of magnetic loops with effectively nothing in between them: where there are no magnetic loops there is no visible corona. Whereas it was previously thought that the magnetic fields acted only to modify the structure of a corona that had certain average properties of temperature and density, there is now a strong body of opinion to suggest that the magnetic fields constitute the basic elements of the corona, and that without the magnetic fields there would indeed be no corona.

**4. Coronal magnetic field**
In this image of the corona, a number of theoretically extrapolated field lines have been plotted over an Hα image superimposed on an eclipse photograph (November 1966).

**5. Coronal loops**
A number of well-formed loops in the inner corona, which were found to correspond with bright prominences seen in Hα light, can be seen in this photograph taken in the light of FeXIV.

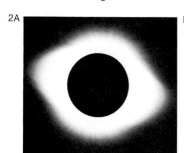

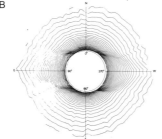

**1. Corona**
The eclipse of 12 November 1966 occurred during a year of relatively high solar activity (1969 was a solar maximum). Coronal streamers can be seen stretching out to a distance of about $4.5\,R_S$; typical "Helmut" streamers lie over the prominences visible on the limb; arch structures are apparent directly above the prominences.

**2. Corona at minimum**
A photograph taken on 30 June 1954 (**A**) shows that the corona is more regular in shape than at maximum, with less fine structure to be seen. The contour lines (**B**) represent equal intensities spaced at intervals of 0.2 magnitude.

**3. Ultraviolet solar images**
The UV images (**B**, **C** and **D**) were taken at wavelengths representing progressively higher temperatures ($10^4$ K, $3.25 \times 10^5$ K, and $2.25 \times 10^6$ K respectively). They thus depict the Sun at successively greater altitudes above the chromosphere, as seen in the Hα image (**A**).

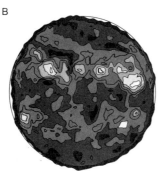

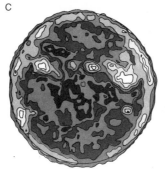

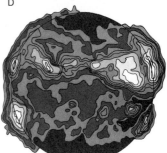

# The Corona II

The optical corona is divided into three principal components: the K-corona, which exhibits a continuous spectrum in which the Fraunhofer lines are invisible; the F-corona, which also exhibits a continuous spectrum of reflected photospheric light but in which the Fraunhofer lines are visible; and the E-corona, consisting of a series of faint emission lines. Overall, the spectrum comprises a faint continuum on which are superimposed weak emission lines.

The K-corona is the inner part of the corona. Its light is photospheric light scattered by very fast-moving electrons, and the absence of Fraunhofer lines is caused by the very high Doppler shifts (*see* page 31) associated with the random velocities of the particles: the Fraunhofer lines of the photospheric spectrum are "smeared" out by these shifts, and are therefore not discernible against the general background light.

The F-corona, or outer corona, is responsible for most of the light beyond a distance of about $2\,R_S$. It comprises relatively slow-moving particles of interplanetary dust, and is really not so much a part of the solar atmosphere as that part of the interplanetary medium which happens to be close to the Sun. Photospheric light is scattered from these dust particles, and because of their relatively slow motions Doppler broadening is insufficient to blur out the Fraunhofer lines. The "white light" corona visible from the Earth at eclipses and from space with coronagraphs comprises the combined contributions of the K- and F-coronas.

The E-corona is composed of light emitted by relatively slow-moving ions (which move more slowly because of their greater masses), giving rise to an emission line spectrum. These lines remained a major mystery to the astronomical community for a long time after the coronal spectrum was first observed in 1869. Particularly prominent among the spectral lines was a line in the green which—like the rest—did not coincide in position with the familiar spectral line associated with any known chemical elements. For a long time it was assumed that the corona contained an unknown chemical element, which was provisionally called "coronium"; after all, the element helium had been discovered in the solar spectrum before it was identified on Earth.

In 1940, however, it was discovered that these lines correspond to transitions that could take place in highly ionized common metals under conditions of very high temperature and very low pressure. These transitions are known as "forbidden lines" because they cannot occur under normal laboratory conditions. The coronal green line at 530.286 nm turned out to be due to 13 times ionized iron (FeXIV). Lines due to highly ionized nickel and calcium also feature prominently. The existence of emission lines due to such highly ionized species provided the first clear evidence that the corona must be at a very high temperature, in excess of $10^6$ K.

**Temperature of the corona**
The temperature of the corona involves a similar problem to that of the chromospheric temperature (*see* pages 34–35): according to the second law of thermodynamics, heat cannot be transferred from a cooler to a hotter body, and it therefore appears impossible at first

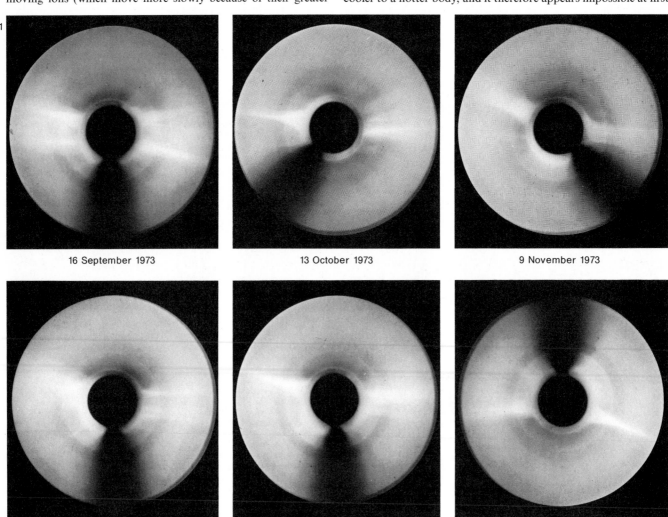

1

| | | |
|---|---|---|
| 16 September 1973 | 13 October 1973 | 9 November 1973 |
| 2 July 1973 | 15 July 1973 | 29 July 1973 |

sight for the corona to be a higher temperature than the 6,000 K photosphere. For a time astronomers remained puzzled by the anomalous figures, but the very high temperature of the corona has been confirmed by a wide variety of observational techniques, including radio observations, the already mentioned absence of Fraunhofer lines in the K-corona, and the presence of highly ionized metals in the E-corona. More detailed measurements of the widths of forbidden lines broadened by the Doppler effect and of the relative intensities of various lines confirm the view that the particles in the corona have a high average level of energy—in other words, the gas has a high temperature. A nominal value for the temperature of the lower K-corona is $2 \times 10^6$ K, although the temperature in different regions of the corona can range from about 1 to $5 \times 10^6$ K. The density of material in the corona, like its temperature, decreases with increasing distance from the Sun: the mean density in the inner corona is (in terms of electron densities) $10^{15}$ m$^{-3}$, as compared with a value some 10,000 times lower at an altitude of about $2 R_S$. Thus although the coronal temperature is very high, and although the energy of each coronal particle is correspondingly high, the particle density is so low that the total amount of energy contained in and emitted by the corona is relatively small. (In fact, the white light intensity of the corona is less than one millionth of the intensity of the photosphere.)

Recently the widely held view of the basic process by which the corona was heated to such high temperatures has been questioned; although the detailed mechanisms were a matter of dispute it was felt that shock waves resulting from the bubbling, noisy convective activity in the photosphere were propagated through the chromosphere and deposited energy in the corona. However, the highest levels of energy radiation are emitted from the regions of strongest closed magnetic structures, and some astronomers have suggested that magnetic effects, rather than mechanical ones, may be the basic source of coronal heating. Two mechanisms are suggested: magnetic field annihilation and MHD waves.

Electrons flowing under the action of coronal magnetic fields experience resistance and so release energy in the form of radiation. In complex twisted magnetic structures—such as may occur when a magnetic flux tube is emerging from the photosphere to form an active region (*see* pages 44–45)—concentrated tubes are readily prone to kinking: lines of force with opposing directions may meet and join up, releasing energy rapidly in the process (as in flares). This mechanism may be responsible for at least some of the energy dissipation in the corona. At the very least, even if the conventional mechanical energy theory is correct, the magnetic field plays the dominant role in channelling the flow of matter and energy through the corona. For the moment, however, it is not certain whether acoustic wave energy or magnetic energy is the primary source of the coronal heating; most probably both processes are at work. Another source of energy is turbulence in prominences generating MHD waves in the corona. It has been suggested that this process may contribute significantly to the heating of the corona in the loops above such prominences.

**1. Coronagraph images**
This sequence shows the development of the corona at intervals of one solar rotation (about 27 days) as seen from Skylab.

**2. Extreme ultraviolet spectrum**
Emission lines from several highly ionized atoms resulting from the high coronal temperatures show up as dark lines in this negative print, in which three different spectral orders can be seen.

**3. EUV and X-ray images**
These simultaneous images portray the Sun in six different chromospheric and coronal wavelengths. The vertical line represents a height of 240,000 km.

**4. Structure of the corona**
Even when viewing conditions are optimum, it is only possible to see a very small portion of the dominant K-corona, out to a distance of about 1.2 $R_S$. During a total eclipse, the E- and F-components may also be seen, out to a distance of 4 $R_S$ or more.

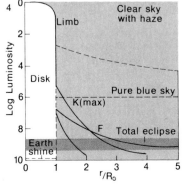

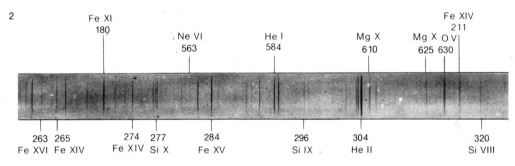

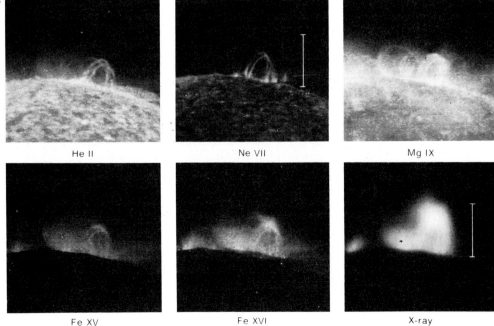

# Coronal Holes and Transients

Coronal holes delineate regions of weak magnetic field in which the field lines are "open"—in other words, they do not curve around to form closed loops connecting regions of opposite polarity. Instead they spread out into interplanetary space, diverging rapidly so that the angular size of a coronal hole increases greatly with increasing distance from the Sun: for example, at a distance of $3 R_S$ the angular area spanned by a hole can be about seven times greater than it is at a level near the base of the corona. Holes show up as dark patches in ultraviolet and X-ray photographs and can cover a large proportion of the visible disc; they are normally present in the polar caps, and it is from these polar coronal holes that the polar plumes emerge, having their bases in bright points.

With temperatures of about $1 \times 10^6$ K they are two to five times cooler than coronal active regions, and cooler, too, than the less bright quiet regions; densities are also correspondingly lower, at about 30 percent of the neighboring bright regions.

Out of coronal holes flow streams of particles, following the field lines and making up the "solar wind" (*see* pages 76–77). These particles emerge from the coronal holes and are accelerated to their maximum velocities within 2 to $3 R_S$ of the surface.

Since the discovery of coronal holes their evolution has been observed in some detail. In outline, the process begins with the appearance of a small isolated hole near the equator. This feature grows in size, rotating with the Sun as if it were attached to a solid surface, and eventually links up with the polar region of the same magnetic polarity. The hole then begins to shrink in area (*see* diagram 1). The entire sequence of events lasts for up to seven or eight months.

## Coronal transients

First studied in detail by the coronagraphs aboard Skylab, coronal transients take the form of giant loops of coronal material hurled out through the corona into interplanetary space. Usually triggered by flares involving mass ejection or, more frequently, by eruptive prominences, these loops or arcs travel outward at speeds of between 200 and $900 \, \text{km s}^{-1}$ (according to Skylab and SMM measurements). Observations made in 1980 close to solar maximum showed that these events spanned a broader range of latitudes (the first 22 events studied spanned 68°N to 54°S) than the Skylab events, which had been observed at a time relatively closer to sunspot minimum (less than 50° latitude). They were also less strongly clustered around the equator; in this they mirror the latitude behavior of phenomena such as coronal extensions, active regions and prominences.

Throughout the duration of the SMM, commencing in 1980, space-borne and ground-based observations were correlated; in particular the SMM coronagraph was used in conjunction with the radioheliograph at Culgoora, Australia, to yield spectacular observations of coronal transients. The loops themselves are relatively thick and structureless and are followed by cavities of less than average coronal density as they move outward. Typical transients of the Skylab era had masses of about $5 \times 10^{12}$ kg and moved at speeds of $470 \, \text{km s}^{-1}$; the energy ultimately contained in this material was about 10 times greater than the radiative energy observed in the initiating flares, and the magnetic energy involved in the restructuring of the field may have been even greater. Thus the total active region energy release associated with flares over

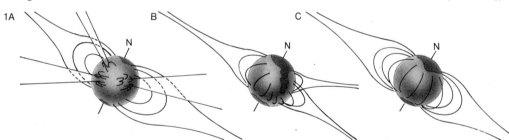

**1. Evolution of a coronal hole**
Small, isolated coronal holes (**A**) appear to link up with one of the polar caps (**B**). Later the hole contracts to the polar region (**C**).

**2. Coronal hole**
These X-ray images showing a coronal hole were taken on 9 and 10 October 1973. A transient X-ray brightening is indicated by an arrow (**A**).

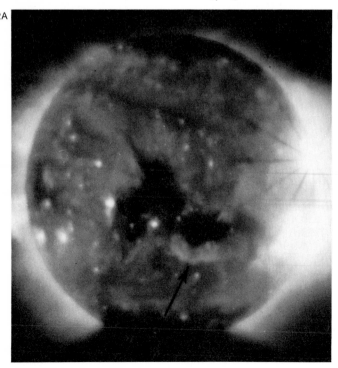

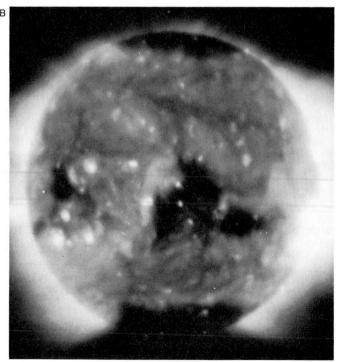

time scales much larger than the flare itself goes mainly into
magnetic and mass motion forms. However, during the flare proper
(triggered by the slow evolution of the active region structure)
power is released more or less equally into radiation and mass
motion. Transients seem to be responsible for major rapid restructuring of the corona. For example, in the case of the event of 7 April
1980, before the transient there was only one streamer visible in the
quadrant over which it occurred, but afterward there were five.

The frequency, scale and violence of these events has led to a
significant modification of the view of the corona and its structure.
Coronal transients are (geometrically) the largest scale impulsive
events so far observed and, since their effects propagate through the
solar wind to the Earth, they can be noticed here.

The corona, then, is seen to be entirely different from the earlier
conception of a relatively smooth, quiet, homogeneous solar
atmosphere. Instead its very existence as well as its detailed and
overall structure has been shown to be in part determined by the
existence of closed magnetic loops that both channel the energy
flow and, possibly, provide a major source of heating. It is entirely
non-homogeneous, having a clumpy structure comprising hot,
bright, relatively dense active regions, cooler, less bright, quiet
regions, and even cooler, faint, rarefied holes from which flows the
solar wind. Finally it is seen to be wracked by powerful transient
events by means of which material is propagated through the
corona at speeds of up to 1,000 km s$^{-1}$ leading to major restructuring. The fact that these events have a high frequency even near solar
minimum implies that they are not tied to the most violent flare
events: rather they tend to be associated with prominence activity.

3

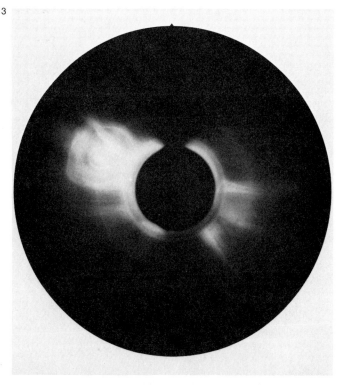

**3. Coronal transient**
This disruption of the corona
occurred on 10 June 1973,
triggered by an eruptive
prominence. The disturbance
travelled rapidly through the
corona, taking less than half an
hour to cover a distance of nearly
a million kilometers.

**4. X-ray images of coronal hole**
This sequence of images shows
the development of a coronal hole
during six solar rotations. Unlike
photospheric and chromospheric
features, coronal holes appear to
rotate as if they were attached to
a solid object.

**5. Polar hole**
In this negative image the polar
hole shows up as a bright "bald"
patch from which polar plumes
may be seen emanating. The
image was made in the light of
MgIX (at a wavelength of
36.8 nm) in the ultraviolet.

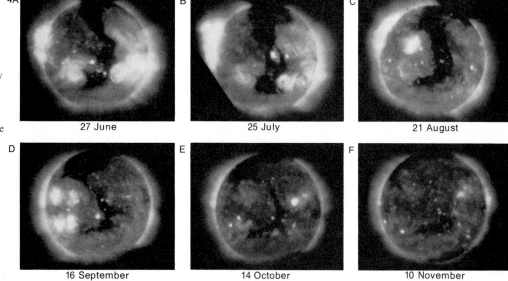

4A          27 June        B        25 July        C        21 August

D          16 September        E        14 October        F        10 November

5

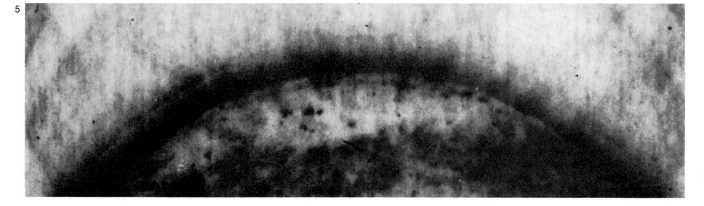

# The Solar Wind

It was suggested in the nineteenth century that the Sun might be the source of transient clouds of plasma that travelled through interplanetary space, impinging on the Earth's atmosphere and giving rise to phenomena such as the aurorae and magnetic storms (disturbances in the Earth's magnetic field). The idea was the subject of a major paper by Sir Oliver Lodge, published in 1900, and in 1932 J. Bartels pointed out that moderate magnetic storms, presumably not caused by violent transient solar flares, had a tendency to recur at 27-day intervals. This interval is the synodic rotation period of the Sun, and Bartels postulated the existence of regions of the Sun that were responsible for producing magnetic disturbances. Such regions came to be known as "M-regions".

Evidence for a more widespread and continuous emission of particles from the Sun came from studies of the tails of comets. By 1958 E. N. Parker had developed a model of the corona that showed that it must be in a constant state of expansion, losing material into the interplanetary medium in all directions in a phenomenon that Parker called the "solar wind". Parker showed that because of the very high temperature of the corona the pressure exerted would ensure an outward flow of material: even at very great distances this pressure could not be contained by any known external pressure. Just as a sand pile will always flow out if its slope exceeds a certain value, so the corona would expand and lose material, which would be replaced from below. Spicules and macrospicules may be major sources of this material from the chromosphere, but the exact mechanism for the supply of matter is not yet clearly understood.

Direct confirmation of the existence of a continuous solar wind came from Soviet and American interplanetary space probes (in particular, Mariner 2) and the solar wind has subsequently been studied *in situ* by a wide variety of spacecraft.

## Properties of the solar wind
The solar wind, although gusty and variable, blows continuously with velocities that can range from as low as 200 km s⁻¹ to as high as nearly 900 km s⁻¹. The typical value is between 400 and 500 km s⁻¹, and the particles normally take about four or five days to reach the Earth. The wind consists of a roughly equal number of electrons and protons, together with a small proportion of heavier ions and nuclei, of which the most prominent are alpha particles (helium nuclei), which comprise, typically, about 4 to 5 percent by number of the total. At the Earth's distance from the Sun, the particle density is on average some $5 \times 10^{-6}$ m⁻³, but it can vary by a factor of over 100. (For comparison the particle density at sea level in the Earth's atmosphere is $2.5 \times 10^{25}$ m⁻³.) The temperature of the plasma is defined by the "velocity dispersion" of the particles—the magnitude of their random velocities relative to the mean flow of the wind. Near the Earth this temperature is on average about $10^5$ K, and the Earth is thus enshrouded by high temperature (but extremely rarefied) plasma. These parameters indicate that the Sun is losing about $10^9$ kg s⁻¹ to the solar wind. At this rate, however, the entire solar mass would take some $6 \times 10^{13}$ years to blow away completely, which is $10^4$ times longer than the Sun will continue in its present form.

## The interplanetary field
Close to the Sun the entire structure is controlled by closed magnetic loops that contain most of the material (*see* page 71), but at larger distances, open magnetic field lines spread out into interplanetary space, somewhat like the spokes of a wheel. Within a few solar radii this magnetic field causes the corona and solar wind plasma to co-rotate with the globe of the Sun. However, beyond a certain distance, the energy of the plasma exceeds that of the field. Although individually the solar wind particles move out more or less radially from the Sun, collectively they form a spiral pattern, in rather the same way as water droplets from a garden sprinkler. Because the solar wind is ionized, it is electrically conductive, and the field is frozen into the plasma. Thus the field lines must also take up the spiral pattern of the particle stream. By about the Earth's distance from the Sun, the field lines of the interplanetary field

**1. Idealized magnetic field**
A simple "dipole" field structure (shown as areas of tone) has been imposed on a simplified model of the corona: the resulting pattern of field lines reveals a "neutral sheet" extending from the equator from a distance of about 2 $R_S$.

**2. Solar wind**
As the Sun rotates, the particles of which the solar wind is composed fan out rather like droplets of water from a garden sprinkler: close to the Sun, however (within the circumscribed region in the diagram), the Sun's magnetic field is sufficiently strong to cause the particles to co-rotate with the Sun as though they were rigidly attached.

**3. Model of magnetic structure**
In these schematic diagrams, the Sun's magnetic structure at the base of the corona is represented as consisting of two components: a pattern of alternating positive and negative polarities near the equator (**A**), and tilted dipole field at the poles (**B**). The two rotate at different rates, and their sum depends on their relative phase.

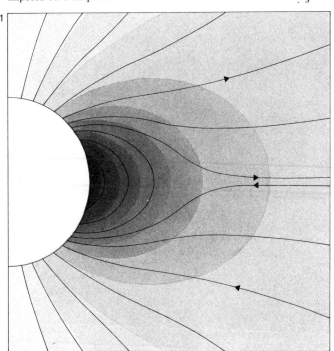

1

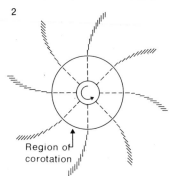

2

Region of corotation

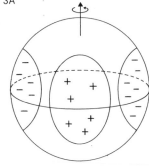

3A

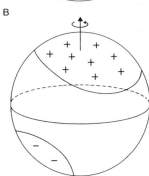

B

make an angle of some 45° with the radial direction from the Sun, while the solar wind particles themselves move almost radially (*see* diagram 2). Observations show that although on the small scale the field lines wave around in a chaotic way, when averaged over a longer period they follow the expected spiral pattern.

Near the solar equator the field is usually found to be divided into two or four sectors of alternate polarity. These sectors are rooted in the large-scale regions of the photosphere, which have a net single polarity (*see* diagram 3). Since the field lines carried by the wind are rooted in the photosphere it is hardly surprising that the interplanetary field should mirror the sector polarities evident in the photosphere.

A consistent view of the nature of these sectors and the source of the high-speed streams is now beginning to emerge. Around sunspot minimum, active regions are concentrated in the equatorial regions of the Sun, and the net polar fields are well established. If the solar field were a simple dipole field it could be represented by closed magnetic structures in the inner corona at the equator, and open field lines emerging from higher latitudes. The open lines would bend over the closed equatorial loops and eventually extend out radially above and below the solar magnetic equator. A "neutral sheet" would be formed across which the direction of the field lines is reversed and this sheet would spread out through the interplanetary medium. However, this simple pattern is disturbed by the sectors of net single polarity in the photosphere, which warp the neutral sheet. At the solar surface the neutral line, instead of following a straight line along the equator, waves to and fro across the equator in a roughly sinusoidal form between heliographic latitudes of around ± 40° (*see* diagram 4). The pattern extends outward, to produce a warped neutral sheet in interplanetary space, and as the Sun rotates the warped boundary passes the Earth at regular intervals, giving an Earth-based observer the impression of sectors of opposite magnetic polarity (depending on whether the neutral sheet passes above or below the Earth). The duration, and hence angular diameter of the sectors, varies in a seasonal fashion as the Earth passes above or below the solar equator.

The magnitude of the seasonal effect indicates that the vertical extent of the distortions is only about ± 0.25 A.U. at the distance of 1 A.U. Thus the spread in latitude of the wobbles may be only about 15° at 1 A.U. compared to 40° at the solar surface. This assertion has been confirmed by space-probes that have moved some distance from the ecliptic. For example, Pioneer 11 (which flew past Jupiter in 1974 and Saturn in 1979) reached 16° above the ecliptic in February 1976 and experienced an interplanetary field with a constant direction for several months. In effect it was for a time above the warped neutral sheet and was experiencing essentially one polarity of an overall dipole field.

The solar wind speed is low over any neutral line or sector boundary and attains high speed away from such a boundary. If the warps in the neutral sheet are sufficiently great, a high-speed stream will be observed when each resulting "sector" passes by. These sectors are often, but not always, filled by a coronal hole. Since large coronal holes are almost always present at each pole, it may be that the major part of the solar wind flows from these holes. Very high-speed streams (750 km s⁻¹) have been observed when the polar coronal holes have extensions towards the solar equator, and it would appear that high-speed streams emanate from polar regions, or from any coronal hole.

The overall effective dipole field—particularly when well established, as it is around sunspot minimum—controls the general form of the interplanetary field and the flow of solar wind particles in such a way that it is likely that the major part of the wind emanates from the polar regions. The high speed of the streams emanating from the polar hole extensions is probably typical of the wind flow over those major parts of the hemispheres away from the vicinity of the warped neutral sheet, above and below 20° or so of the solar equatorial plane. The Out-Of-The-Ecliptic/Solar Polar mission, scheduled for the early 1980s, will travel via Jupiter before doubling back over one of the Sun's poles, and should greatly clarify a phenomenon which is so far unexplored.

**4. Coronal brightness maps**
In an idealized situation (**A**), the Sun's tilted dipole field would give rise to a sinusoidal wave pattern in the neutral line. (**B**) is an actual example showing the coronal brightness as measured for Carrington rotation 1602.

**5. Three-dimensional model**
The Sun's magnetic and density structure for Carrington rotation 1602 (the same as diagram 4B) is illustrated as it might be imagined in three dimensions. The neutral sheet is distorted in accordance with the inferred pattern.

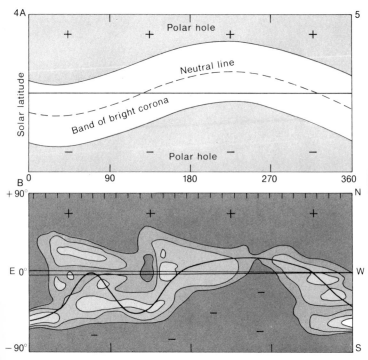

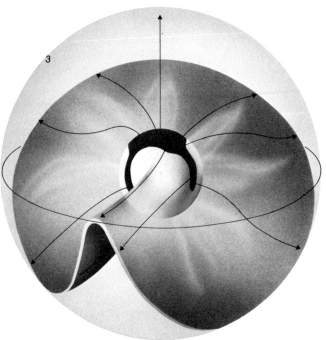

# Solar Interaction with the Earth I

Apart from providing light and heat the Sun interacts with the Earth in a wide variety of more subtle ways. Its influence on the Earth's magnetic field, for example, is considerable. Like the solar field, the geomagnetic field may be represented (to a first approximation) as a simple dipole, with a mean strength at the poles of about 0.6 G; the position of the Earth's magnetic poles alters slowly with time. The region in space within which the Earth's field is the predominant magnetic influence is called the "magnetosphere", and if it were not for the effect of the solar wind, this region would extend out to a distance of about 100 Earth radii ($R_E$), affecting the motion of charged particles within this region. However, the pressure of the solar wind compresses the sunward side of the magnetosphere to within 8 to 10 $R_E$, while in the opposite direction, interaction with the wind draws out the field lines into a "magnetotail", stretching well beyond the orbit of the Moon. The magnetotail has been detected as far as 1,000 $R_E$ from the planet. It contains a "neutral sheet", on either side of which the field lines have opposite direction.

The boundary of the magnetosphere, across which the solar wind cannot readily flow, is called the "magnetopause". The solar wind flows much faster (about eight times faster on average) than the speed at which sound waves can pass through it, and the magnetosphere can thus be regarded as effectively "moving" supersonically through the wind. Just as a supersonic aircraft will pile up material in front of it forming a shock wave, so the magnetosphere produces a shock wave in the solar wind called the "bow shock". On the sunward side this bow wave precedes the magnetopause by about 3 or 4 $R_E$. At the shock front, the speeds of solar wind particles are abruptly slowed from about 400 km s$^{-1}$ to about 250 km s$^{-1}$, and the kinetic energy of motion lost in this process is transformed into heat, raising the temperature of the plasma to several million K, 5 to 10 times higher than normal for the wind. Between the bow shock and the magnetopause is a region

known as the "magnetosheath", characterized by jumbled and chaotic magnetic field lines.

## Magnetic storms

Variations in the solar wind produce changes in the magnetosphere that are reflected in the terrestrial magnetic field at ground level in phenomena known as magnetic storms. The bulk of solar flare particles reach the Earth after about 2 days from the occurrence of the flare; the shock wave preceding the cloud of plasma compresses the magnetosphere, rapidly intensifying the geomagnetic field at ground level. This phase, which takes place over a timescale of a few minutes, is called the "sudden storm commencement" (SSC). It is followed by the "initial phase" (IP), lasting from about 30 min to a few hours, during which time the flare plasma flows past the Earth while the field strength remains higher than it was before SSC. During the next phase, the "main phase" (MP), the particle population is enhanced and particles are accelerated by the release of magnetic energy, particularly in the magnetotail where field lines reconnect. These effects cause a flow of current in the magnetosphere, which itself generates a field opposed to that of the Earth; as a result there is a drop in the geomagnetic field strength lasting between a few hours and about a day. As the current decays, the field strength returns to normal, possibly over a few days. Magnetic storms of this kind may recur at 27-day intervals where the source is a persistent active region.

## Van Allen belts

There are certain restricted regions in the Earth's magnetosphere within which highly energetic charged particles are trapped by the magnetic field. These regions, which were discovered in 1958 by the space-probes Explorers 1 and 3, are known as the "Van Allen belts". Conventionally two zones are shown, at about 1.5 and 5 $R_E$ from the center of the Earth, but there is no clear-cut division;

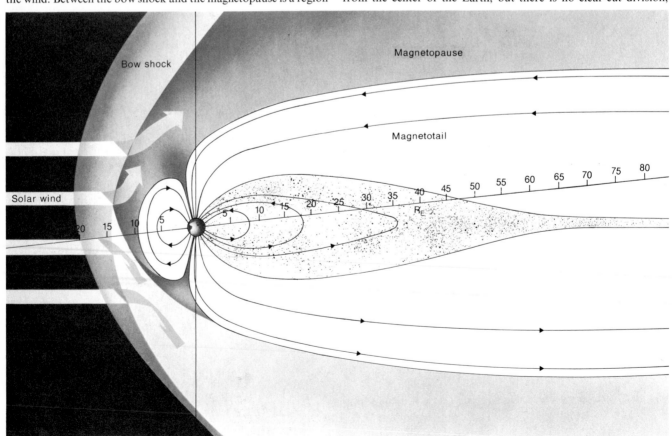

however, the inner belt has a higher concentration of high-energy protons and electrons than the outer belt. The charged particles in these belts are bound to the magnetic field and spiral to and fro along the north–south lines, since they are reflected back when they meet the converging lines near the poles; the period of one to-and-fro oscillation is of the order of 1 sec. The population of particles in these belts is maintained by the solar wind, by cosmic rays and by particles sprayed from the upper atmosphere by cosmic ray impacts, combined with acceleration of these particles in the Earth's field by processes similar to those occurring in solar flares.

### The ionosphere

Solar X-ray and ultraviolet radiation photoionizes atoms and "photodissociates" molecules in the Earth's upper atmosphere resulting in the presence of free electrons stripped from their atoms and molecules. These have a tendency to recombine, but new ions are created at an equal rate, and at any one time the concentration of ions is sufficient to cause variety of phenomena. A number of regions of ionization in the atmosphere are recognized, each owing its level of ionization to different wavelengths of radiation. Of particular interest is the D-region, extending up to 90 km above ground level, where ionization is due to ultraviolet radiation (notably $Ly_\alpha$), extreme ultraviolet, X-rays and cosmic rays. This layer reflects long-wave radio waves in the wavelength range of 10 to 100 m allowing radio signals to be used for communication around the globe, but preventing ground-based long-wave radio astronomy. While the ionosphere reflects long wavelengths, it also absorbs very short wavelengths, and so prevents the penetration to ground level of a range of frequencies originating in space. This absorption also gives rise to enhanced levels of ionization which occur after solar flares and related events because of an increased incidence of X-ray radiation. As a result, some radio frequencies are sometimes absorbed and the signal may fade.

Much more drastic ionospheric disturbances are caused by violent flares called "solar proton events", in which copious quantities of higher energy protons and other nuclei are ejected with energies of tens or hundreds of mega electron volts (MeV). These particles penetrate the barrier normally provided by the magnetosphere and enter the Earth's atmosphere by spiralling along the field lines at high geomagnetic latitudes. They can produce sufficient ionization to black out radio signals and severely attenuate cosmic radio waves for periods lasting a few days. The frequency of these Polar Cap Absorptions (PCA) correlates well with the solar cycle: 10 or more events may occur in a year near maximum, while near solar minimum there may be none at all.

### The ozone layer

Solar ultraviolet radiation impinges on the upper atmosphere and breaks down molecular oxygen ($O_2$) into separate oxygen atoms. In the atmosphere between 15 and 50 km altitude, ozone ($O_3$) is formed by a combination of O and $O_2$. Ozone is destroyed by ultraviolet radiation, particularly in the 210 to 310 nm range. In this process it heavily absorbs the radiation, which is harmful to living tissue. Equilibrium is normally maintained between creation and destruction of ozone. The peak concentration of ozone tends to occur at an altitude of about 20 km, a region known as the "ozone layer", but even there its concentration is only about 10 parts per million. Nevertheless this thin concentration is essential for the preservation of human life on this planet.

Another of the consequences of the enhanced ionization brought about by a solar proton event is the production of atomic and ionized nitrogen, which reacts with $O_2$ and ozone to produce nitric oxide (NO), depleting the quantities of ozone. After the massive proton flare of 4 August 1972 it was reported that there was a 16 percent decrease in ozone over high altitudes, with consequent increased penetration to ground level of ultraviolet radiation.

**1. Earth's magnetosphere**
On its sunward side the Earth's magnetosphere is compressed by the highly energetic particles of the solar wind, which collide with the Earth's magnetic field. On the opposite side, the "magnetotail" stretches far out into space.

**2. Van Allen belts**
The Earth is encircled by zones of radiation known as "Van Allen belts", in which charged particles spiral to and fro, trapped by the Earth's magnetic field; the belts are therefore inclined at an angle to the Earth's rotational axis.

**3. Ionosphere and ozone layers**
X-ray and ultraviolet radiation from the Sun ionizes atoms in the upper regions of the Earth's atmosphere, producing a layer known as the "ionosphere". This layer reflects certain long-wave radio signals. Variations in solar activity alter the concentration of ions in this region, and may possibly influence the frequency or intensity of thunderstorms. At a lower altitude (about 20 to 30 km) lies the "ozone layer", which is also the product of solar ultraviolet radiation.

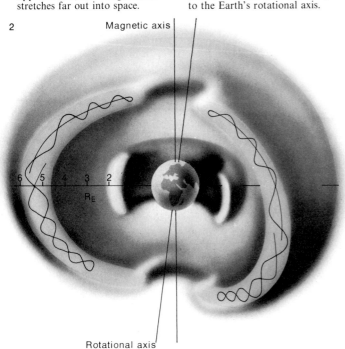

2 Magnetic axis

Rotational axis

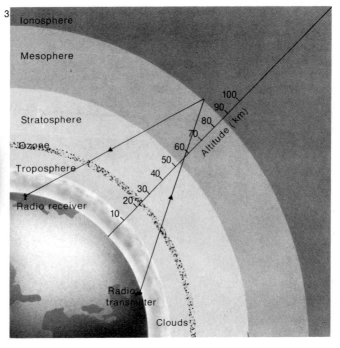

3 Ionosphere

Mesosphere

Stratosphere

Ozone

Troposphere

Radio receiver

Altitude (km)

100
90
80
70
60
50
40
30
20
10

Radio transmitter

Clouds

# Solar Interaction with the Earth II

Many attempts have been made to correlate specific terrestrial phenomena with the level of solar activity, some more bizarre than others—ranging from the price of wheat and the quality of wine to political upheavals. Most of these so-called correlations are spurious. Nevertheless there is now mounting evidence to show that variations in solar activity are reflected in atmospheric phenomena, weather and, perhaps, long-term climatic change. For example, one of the better established associations is that the passage of sector boundaries in the interplanetary field—which typically takes place every 7 or 14 days—has been shown to be correlated with changes in the amount of atmospheric "vorticity" (the amount of rotation present in cyclonic phenomena). Vorticity initially decreases in the first day after the passage of a sector boundary, then rises to a maximum $3\frac{1}{2}$ days after the passage of the boundary. Polar atmospheric pressure increases and middle latitude pressures decrease after the boundary passage, with the result that the stronger anticyclones are established in higher latitudes and greater storminess is evident in lower latitudes.

Such atmospheric effects, however, are very far from proven. In contrast, one undoubted influence of solar activity on the Earth's upper atmosphere is the effect of solar particles, emitted by solar flares, on high-latitude regions, where they give rise to aurorae. Aurorae are colored patterns or rays and arcs, predominantly red and green in color, which are periodically seen in the sky from the extreme northerly and southerly parts of the Earth. The green color is due to the emission of the green oxygen line at a wavelength of 557.7 nm. The aurorae are caused by excitation and ionization of the upper atmosphere by influx of electrons, probably accelerated towards the Earth by disturbances in the magnetotail. Following the geometry of the geomagnetic field, particles enter the upper atmosphere primarily within a thin oval band lying between 10 and 20 degrees of the magnetic poles. Around the time of sunspot maximum the auroral band migrates towards the equator, allowing displays to be seen from lower latitudes.

The frequency of auroral displays is closely linked to the solar cycle, and historical auroral records (or rather, the lack of them)

provide useful corroborative evidence for the reality of the Maunder Minimum (*see* pages 40–41). Under present-day conditions between 500 and 1,000 aurorae should be visible in populated areas of Europe during a 70-year period, and yet between 1645 and 1715 there are scarcely any reports; for one spell of 37 years there was not one report. Reports of aurorae increased rapidly in 1550, were interrupted by the minimum and then jumped by a factor of 20 after 1716. There were no reports of naked-eye sightings during the same period.

## The carbon-14 record

One interesting suggestion, which admittedly has been strongly criticized recently, is that the radioactive carbon-14 found in the growth rings of trees provides indirect evidence of long-term variations of the Sun's output of radiation. Carbon-14 is produced by cosmic rays in the upper atmosphere, and the intensity of cosmic rays is modulated by the solar cycle. It is believed that when solar activity is high, the interplanetary magnetic field is more effective at scattering cosmic rays away from the Solar System and the observed level of cosmic rays is reduced; at times of low activity the cosmic ray flux is enhanced. Carbon-14 is absorbed into tree rings during each year's growth cycle, and the ratio of carbon-14 to carbon-12 (common carbon) in the rings can be determined by analysis. Measurements of the carbon-14 content of tree rings dating from the Maunder Minimum have been exceptionally high, in accordance with the hypothesis that solar activity was virtually absent, and that the interplanetary magnetic field was therefore weaker and less complicated.

The tree ring record also shows another period of depressed solar activity between about 1450 and 1550, a period now named the "Spörer minimum" in honor of the man whose work drew Maunder's attention to the anomalous sunspot records. Moreover, abnormally *low* levels of carbon-14 characterize the period from about 1100 to 1300—the medieval maximum. Sparse corroborative evidence comes from early Chinese, Japanese and Korean records of naked-eye sunspots, dating back to at least 28 BC. Auroral

1

## 1. Aurora
This view of the southern aurora was taken from Skylab and shows the bright patterns near the edge of the auroral cap; the permanent aurora over the south pole can be seen in the background. Skylab was moving into the sunlight when the photograph was taken.

## 2. Auroral records
Historical evidence based on the frequency of reported sightings of the northern lights appears to confirm the existence of the "Maunder Minimum", lasting from 1645 until 1715. The virtual absence of auroral reports (black) during this period accords with the similar absence of sunspot records (grey). The modern sunspot cycle seems to have begun in the early eighteenth century, and this period also marks a dramatic jump in auroral sightings. Black dots in the graph represent Oriental records of sunspots; these too appear to correspond with a relatively high level of auroral activity.

## 3. Electron path
Aurorae are produced by electrons spiralling in towards the Earth's poles, following the pattern of magnetic field lines.

sightings follow a similar pattern of gaps, although there does seem to have been a massive increase in sightings subsequent to 1716 (the end of the Maunder Minimum and the establishment of the sunspot cycle as it is known today).

### The solar constant
It has been suggested that variations in the total output of quiet solar radiation (the solar "constant") appear to have had a significant effect on the Earth's climate, and if so such variation may similarly affect it in the future. Until very recently direct measurement of the solar constant could not be achieved with great accuracy: ground-based measurements were hopelessly confused by effects of variable atmospheric absorption and only during the past decade has it been possible to make reasonably precise measurements by means of space-borne radiometers, which measure radiation over virtually the entire spectrum from X-rays to radio waves. However, recent investigations have begun to reveal suggestive evidence for long-term variations in the level of solar activity as well as changes in the solar constant. It has been surmised that such slow changes in the energy output generated deep in the solar interior would alter the pattern of convection within the Sun, and this in turn would modify the solar dynamo, thus altering the level of sunspot activity.

The most precise measurements of the solar constant so far have been carried out by the SMM spacecraft (*see* page 24); the mean value for the solar constant obtained by the instrument over the first 153 days of operation was 1,368 W m$^{-2}$ with fluctuations of $\pm 0.05$ percent. Variations in the solar constant of about 0.1 percent have been detected over periods of a few days and two large decreases of 0.2 percent appear to be correlated with the presence of sunspot groups; shorter term variations over minutes or a few hours with values of at least 0.05 percent have also been noted.

In 1976 and 1978 rocket observations of the solar constant showed that over a 30-month period it had increased in value by about 0.4 percent. A third rocket experiment in 1980 indicated that the rise has been sustained at almost the same level. These results are confirmed by observations made by the Nimbus 7 satellite between late 1978 and mid-1979 which indicate a mean value of solar constant about 0.15 percent higher than the 1978 rocket value, and 0.66 percent higher than the 1976 value, consistent with the general trend. It must be emphasized, however, that solar constant radiometers pose particularly difficult problems of calibration.

Although these fluctuations are quite small, they demonstrate clearly that the solar output is not absolutely constant. It has been estimated that systematic changes in solar luminosity of only 0.5 to 1 percent per century could have profound climatic effects and could readily account for the great ice ages of the past as well as shorter term climatic fluctuations. In particular, a decrease of 2.5 percent in the solar constant would move the Earth into a condition of permanent ice age, since the high reflectivity of the resulting ice-cover would prevent the ice from thawing even if the solar constant then returned to its present value. Variability in the solar constant is therefore of considerable interest, although other processes are also equally feasible sources of ice ages.

There is, finally, some evidence of another kind that favors changes in the solar constant linked with the solar cycle as well as, possible, longer term changes. Observations carried out at Lowell Observatory since 1966 have shown closely correlated changes in the observed brightness of the planets Uranus and Neptune. If the albedos of these planets are assumed to have remained unchanged, then these observations are consistent with a variation of up to 2 percent in the solar constant over the 11-year cycle. However, as already mentioned, such a large variation would have had a catastrophic effect on the Earth's climate. In some way the planets appear to be magnifying solar variations. On the other hand, observations of Titan, Uranus and Neptune since 1972 reveal a steady brightening, and more recent observations of the satellites of Jupiter and Saturn are consistent with the idea that solar output has increased by about 0.4 percent since 1970. It is still a matter of debate whether these results indicate a true change in the solar constant or are indicative of an effect produced by solar emissions, altering the albedos of planetary bodies in step with the cycle.

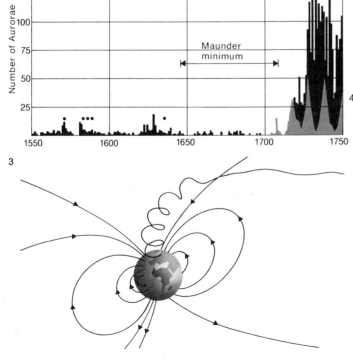

**4. Records of solar activity**
The top curve represents the relative abundance of carbon-14 to carbon-12 in the growth rings of the Bristlecone pines. Increases in the proportion of carbon-14 are plotted as downward shifts in the curve and represent supposed decreases in solar activity. The curve below has been smoothed out to produce a long-term "envelope" of a possible sunspot cycle—the limits of successive maxima and minima. The remaining curves are estimates of past climate, based on the mean winter temperature in England, a winter severity index for the Paris–London area, and the retreat and advance of Alpine glaciers. The correlation is suggestive but not conclusive.

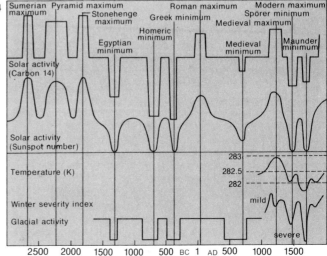

# Solar Energy

Over the years solar physicists have built up a convincing model of the interior of the Sun and the mechanism by which it produces its energy. According to this "standard model", the density, temperature and pressure increase sharply towards the center of the Sun, where the temperature is estimated to be about 15 million K; the density is estimated at $1.58 \times 10^5 \, \text{kg m}^{-3}$ (about 12 times the density of lead) and the pressure $2.5 \times 10^{11}$ Earth atmospheres. The core is believed to extend out to about $0.25 R_S$ and, although this comprises only one sixty-fourth of the Sun's volume, it is believed to contain about 50 percent of the total mass; within this zone about 99 percent of all solar energy is generated. However, most (50 percent) of this energy is generated in only 5 or 10 percent of the solar mass. Despite the very high pressure, the intensely high temperature ensures that the central matter is kept in a fully ionized gaseous state, comprised mainly of protons and electrons moving about at very high speeds. A state of "hydrostatic" equilibrium exists, in which the Sun's tendency to collapse under its own gravitational self-attraction is balanced by the outward pressure of the hot gas inside.

It is assumed that originally the whole of the Sun had the same chemical composition as is now evident in its outer layers, with proportions (by mass) of about 78 percent hydrogen, 20 percent helium and 2 percent other elements (principally carbon, nitrogen and oxygen—see pages 14–15), and that the interior has been modified by thermonuclear reactions so that now the fraction of hydrogen in the core is 36 percent, about half the value at the surface. Beyond about $0.25 R_S$ the composition is believed to be essentially uniform and equal to the surface value.

### The proton-proton chain

The reaction believed to be principally responsible for the production of energy in the solar core is the "proton-proton" reaction or "p-p reaction" (see diagram 1). In a series of stages, four hydrogen nuclei (protons) are fused together to form one helium nucleus, comprising two protons and two neutrons. The helium-4, once formed, is stable because the temperature of the core is too low for the next stage of thermonuclear reaction, involving carbon, to take place. The mass of the end product is 0.7 percent less than the

components that went to assemble it: this small percentage loss of mass is converted to energy. The well-known relationship between mass (m) and energy (E), derived from the Special Theory of Relativity, is $E = mc^2$, where c is the speed of light. Since c is a large number, a very large amount of energy can be released by the destruction of small quantities of matter. In order to sustain the present luminosity just over $4 \times 10^6$ tonnes of matter must be converted into energy every second by this process.

Several different variations can occur in the reaction, but 95 percent of the helium nuclei are produced by the basic p-p chain.

### The solar neutrino problem

Although there is no real doubt about the basic process that sustains the Sun's output—the fusion of hydrogen to form helium—serious doubts have recently been raised about the precise chain involved. Two of the principal strands of observational evidence responsible for these doubts concern the observed flux of neutrinos, and the solar oscillations (see pages 84–85).

The first step in the proton-proton reaction releases neutrinos. Neutrinos are a common type of "particle" (symbol $v$) with zero electrical charge and (it is usually assumed) zero rest mass. Neutrinos travel at the speed of light and very rarely interact with matter, passing through solid objects almost as if they were transparent. Thus neutrinos created at the Sun's core are able to emerge from the surface in a fraction of a second. The neutrinos produced in the p-p reaction turn out to be of rather low energy, usually less than 0.42 MeV. In the comparatively rare alternative routes of the reaction chain higher energy neutrinos are released, particularly the very rare route in which beryllium is converted into boron before breaking down to form helium-4. The beryllium–boron chain occurs only in about 0.01 percent of reactions.

The Sun, then, should be emitting neutrinos, most of rather low energy, with a small proportion of high-energy neutrinos. The rate of their production depends sensitively on the internal temperature of the solar core, so that, if it were possible to measure the flux of neutrinos emerging from the Sun, this would provide a measure of the internal temperature. However, since neutrinos can penetrate the Sun so easily, it is evident that they are extremely difficult to

**1. Proton-proton chain**
The main source of the Sun's energy is believed to be a chain of nuclear reactions in which hydrogen nuclei (protons) are converted to give helium (an atom containing two neutrons and two protons). The simplest route involves only three steps (**A-B-C**). Two protons combine to give a deuterium nucleus (a proton and a neutron), which emits a positron and a neutrino (**A**); the deuterium then combines with a proton, yielding a nucleus of helium-3,

which emits a photon (**B**); finally the helium-3 combines with another helium-3 nucleus (formed in the same way), producing a nucleus of ordinary helium-4 plus two extra protons (**C**). However, several variations in this chain are possible (**A-B-D-E-F**, or **A-B-D-G-H-I**). In these rarer alternative routes, involving beryllium-7, lithium-7 and boron-8, neutrinos with different energies are emitted. The neutrinos given off by boron-8 (**H**) should be detectable by the Brookhaven instrument.

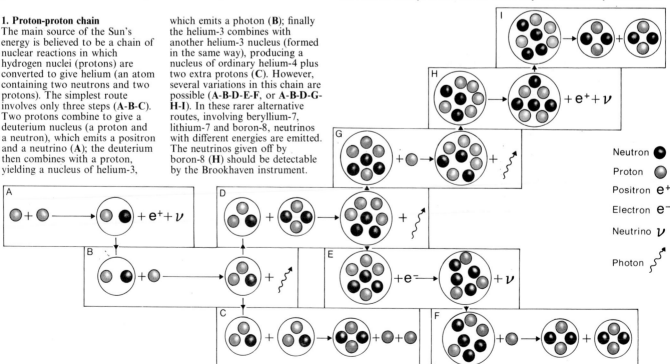

| | |
|---|---|
| Neutron | ● |
| Proton | ● |
| Positron | $e^+$ |
| Electron | $e^-$ |
| Neutrino | $v$ |
| Photon | 〜 |

capture and measure. Nevertheless, in 1964 a neutrino "telescope" was established by the Brookhaven National Laboratory capable of detecting solar neutrinos. The instrument consisted of some 400,000 liters of perchloroethylene ($C_2Cl_4$)—dry-cleaning fluid—containing large amounts of chlorine in a tank located at the bottom of a mine in South Dakota, at a depth of some 1,500 m. The isotope chlorine-37, which comprises about 25 percent of the chlorine content of the dry-cleaning fluid, is effective at capturing the energetic neutrinos. In the process it is converted into radioactive argon (and an electron is given off). The radioactive argon is flushed from the tank by means of helium, and collected in a charcoal trap cooled to the temperature of liquid nitrogen (77 K). The decay of the argon is registered on particle detectors and the number of argon atoms is measured. Thus the flux of neutrinos entering the detector can be determined.

The chance of an individual chlorine-37 atom capturing a neutrino is very tiny indeed: if standard theory is correct, a given chlorine-37 atom in the detector should capture a neutrino once in about $10^{28}$ years—far in excess of the estimated age of the universe. However, the container holds a very large number of atoms. Neutrino astronomers have invented a unit of measurement known as the solar neutrino unit (SNU), where $1\,SNU = 10^{-36}$ neutrino captures per second per target atom. According to the standard model, the detector should measure between 6 and 7 SNU, equivalent to about 1 neutrino capture a day. In the 15 or so years that the experiment has been operating it has failed to detect anything like the expected number of neutrinos. The measured flux is 2.2 SNU, about one third of the predicted value.

If the temperature of the Sun's core were about 10 percent lower than implied by the standard model it would explain the low flux of neutrinos; on the other hand, a lower core temperature would reduce the Sun's luminosity to well below the observed value—unless the Sun is also less opaque than is supposed. But a lower opacity would in turn imply a lower proportion of heavy elements in the outer region, and this would be in conflict with the estimated age of the Sun. It has been suggested that the outer layers of the Sun were accreted from interstellar materials, which might account for a difference in composition between the interior and the exterior.

Another explanation is that the output of energy from the core fluctuates over long periods of time, and that the present level of photospheric radiation characterizes an earlier era when the core was more energetic (photons taking $10^7$ years to escape) while the neutrino flux represents the present level of core activity. More erratically it has been conjectured that a tiny "black hole" resides in the solar interior, digesting material and contributing about half of the radiation output.

A different approach to the problem is to argue that it is the nuclear theory that is at fault. On the one hand, there is a degree of uncertainty about the rate of reactions involving boron; on the other, it may be that more needs to be known about the neutrinos themselves. It has been argued that if the neutrino had a tiny but finite mass it would oscillate between two or more states: there are known to be two types of neutrino ("electron family" neutrinos and "muon family" neutrinos), and there is good reason to suppose that a third type ("tau-neutrinos") also exists. If the neutrinos produced by the decay of boron are distributed fairly evenly between the three types, then the expected rate of detection would be reduced by one third, since chlorine-37 is only capable of detecting electron-family neutrinos. More refined calculations indicate that this effect would only reduce the expected flux by half.

What is needed is a form of neutrino detector capable of picking up the much more common low-energy neutrinos produced in the first stage of the p-p chain. The element gallium appears to offer one of the best possibilities. Plans are afoot to build a gallium detector of some 50 tonnes in the Homestead mine, but funding is a problem: 50 tonnes of gallium—well in excess of annual world production—would cost some 50 million dollars. A 20-tonne gallium detector is under development. Meanwhile the Soviet Union has established a neutrino observatory deep under the Caucasus mountains. A neutrino detector filled with 330 tonnes of white spirit is already complete and should be capable of improving sensitivity over the Brookhaven detector by a factor of about 5. Indium-115 provides another viable detection agent; about 4 tonnes of indium would suffice to achieve a detection of about 1 neutrino per day over the whole energy range. The cost would be between 10 and 20 million dollars.

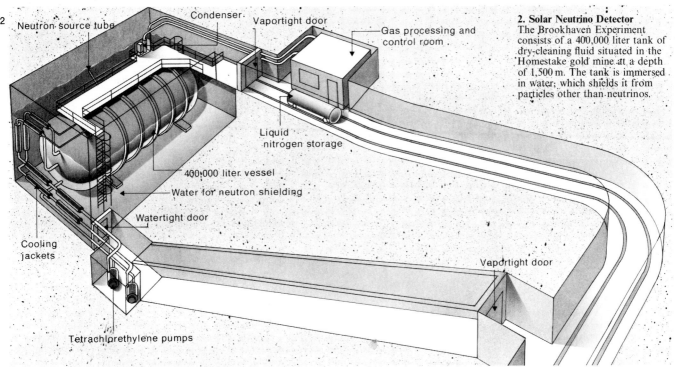

2

Neutron source tube
Condenser
Vaportight door
Gas processing and control room
Liquid nitrogen storage
400,000 liter vessel
Water for neutron shielding
Watertight door
Cooling jackets
Tetrachlorethylene pumps
Vaportight door

**2. Solar Neutrino Detector**
The Brookhaven Experiment consists of a 400,000 liter tank of dry-cleaning fluid situated in the Homestake gold mine at a depth of 1,500 m. The tank is immersed in water, which shields it from particles other than neutrinos.

# Solar Oscillations

In 1973, in the process of making very precise measurements of the ratio of the polar to equatorial diameters of the Sun, R.H. Dicke discovered that the Sun is oscillating, vibrating like a jelly, in such a way that the equator bulges as the poles are flattened, and the poles are elongated as the equator contracts (*see* diagram 1). Since that time it has been established that the Sun oscillates with a variety of periods from about 5 min to just over 1 hr. The difficulty in making such observations can hardly be overstressed: the maximum amplitude (vertical motion) of the solar surface is only about 5 km and maximum velocity about 10 m s$^{-1}$. There are many stars, including conspicuously variable stars such as Cepheids and Long-period variables, that oscillate in a much more pronounced fashion, and this phenomenon was well established in these examples long before the discovery of solar oscillations.

It is not very surprising that the Sun should oscillate in this way. Just as a bell has a natural frequency of oscillation at which it will tend to vibrate when it is struck, so the Sun has natural frequencies at which it will vibrate given a suitable "hammer". A natural oscillation like a bell's vibration dies away due to internal damping unless it is struck periodically in order to maintain its vibration. Likewise, within the Sun there must be a mechanism that keeps the oscillations going: it is assumed that fluctuating nuclear processes in the core are in some way responsible.

Global oscillations of the Sun offer astronomers a means of studying the structure of the solar interior, just as seismic waves from phenomena such as earthquakes have made it possible to determine the interior structure of the Earth. By analyzing the frequencies of solar oscillations attempts are being made to achieve an equivalent understanding of the Sun's interior. Although the range of oscillation periods is broadly consistent with the standard model, there appear to be discrepancies; for example it appears that the range of frequencies is not consistent with the value for the (assumed) depth of the convective zone.

The 5 min oscillation had been discovered as early as 1960, and is generally believed to be a surface ripple affecting only the outermost 10,000 km or so of the solar globe. The other frequencies that have been measured are believed to represent the fundamental radial oscillation of the entire Sun and its harmonics. If the standard model of the interior is correct, then a fundamental period of about 1 hr would be expected and, on the face of it, the observed results seem roughly consistent with the way the Sun's interior is believed to be constructed. However, it has been pointed out that if the oscillations arise in the deep interior, then—because of damping mechanisms—the oscillation seen at the surface should be weaker than those in the interior. Attempts to calculate the oscillation magnitudes towards the core required to match the observations appear to indicate that they would become of such great amplitude that they would disrupt the solar interior.

In 1976 a Soviet research team and a group from Birmingham, England, independently announced the discovery of a long-period global oscillation of 2 hr 40 min. These measurements were not direct measurements of the motion of the solar limb but spectroscopic measurements of selected spectral lines. The Doppler shifts in these lines were measured by comparison with laboratory sources, and periodic variations were interpreted as oscillations of the solar surface as it alternately approached and receded from the observer. If this long-period oscillation were a true fundamental period, then the standard model could not be correct: to account for such a period, the Sun would have to be more homogeneous throughout, and the density could not increase so sharply towards the center as proposed by the standard model. Moreover, the core would be too cool to account for the Sun's luminosity by the proton-proton reaction. Indeed, it appears that if the Sun was shining by means of the p-p reaction *and* if the 2 hr 40 min period was the fundamental oscillation, then the central temperature would be less than half the conventional value, and the luminosity would be less than one ten-thousandth of the observed value.

It was suggested that the 2 hr 40 min period was not an oscillation at all but an apparent effect induced by supergranulation. As a result of the rotation of the Sun, supergranular cells of 15 to 30 thousand km diameter would cross the field of view in periods of between 2 and 4 hr. Since these cells contain rising gas, which

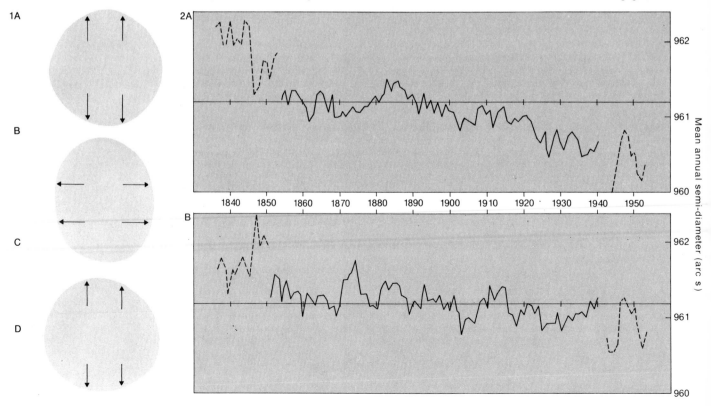

spreads out and falls back to the interior, a significant number of cells moving across the strip of Sun studied by the Soviet instrument would introduce an apparent periodic Doppler shift. However, it does not appear to be possible to explain away the long-period oscillation in this way, and it has been confirmed by other groups since then; there is even suggested evidence for a small periodic variation in the geomagnetic field consistent with the propagation of the surface disturbances by means of the solar wind. Another objection raised was that the observed variations might result from the fact that the distance travelled through the Earth's atmosphere by light from the Sun changes with the Sun's varying altitude in the sky. This possibility seems to be eliminated by the work of a French group at the South Pole, where the solar altitude is near constant. Further checks will soon be made by simultaneous observations from remote sites. It seems certain there is some periodic effect to explain, but whether the oscillation is a true global oscillation or a surface effect, or a "gravity wave" like waves in the ocean, remains a matter of debate.

More recent analyses of the 2 hr 40 min oscillation suggest that it could be accounted for by reducing the core temperature by about 10 percent. This would leave a major problem in accounting for the observed solar luminosity, but it is interesting to note that that same degree of reduction of core temperature would probably suffice to account for the low level of neutrino flux.

### Is the Sun shrinking?
In 1979 evidence was put forward indicating that the Sun was shrinking. From an analysis of solar diameter measurements made at the Royal Observatory, Greenwich, over a period of about 120 years from 1836 to 1954, it was suggested that the diameter was decreasing by about 0.1 percent per century. If this figure were correct, and represented a uniform rate of decrease, then the Sun would have been twice its present size about 100,000 years ago, and would shrink to a point in the next 100,000 years. Such a conclusion is patently absurd, but it was suggested that the Sun expands and contracts periodically, possibly over hundreds of years. If the rate of

energy generation at the core (the fundamental driving force behind these variations) was to fluctuate, then, it was argued, the gravitational energy released in the shrinkage would supplement the energy released from the core. Given the present rate of shrinkage, the extra energy would account for a lower central temperature and would fit in with the low observed neutrino flux.

The validity of the results has been challenged by a number of investigators. During the period spanned by the Greenwich observations both the telescope used and the chronometer were changed. Furthermore, increasing atmospheric pollution at Greenwich over the period would affect the brilliancy of the solar limb and further contribute to an apparent shrinkage.

Different types of evidence were also adduced. For example, the eclipse of 1567 seen from Rome was not fully total, the Moon's disc being surrounded by a ring of light: it was argued that the Sun must have been larger then than at present to account for this. However, detailed analysis of the orientation of the Moon on that date shows that the phenomenon of "Baily's beads" (due to light shining through valleys at the lunar limb) would have ensured that the eclipse in question could not have appeared total. Similarly analysis of the duration of transits of Mercury over the past 250 years and of the duration of eclipses over a similar period appear to show quite clearly that the solar diameter has not altered significantly over that time. However, although it would appear that the Sun is not shrinking to the extent that was first suggested, the matter is still not settled, and the possibility of small-scale periodic changes in solar diameter—linked perhaps to small changes in the solar constant—certainly cannot be ruled out.

Understanding of the solar interior is in a state of flux. No-one seriously doubts that the Sun shines by means of thermonuclear reactions converting hydrogen to helium, but the precise mechanism is open to doubt. The Sun has been shown to be variable, although only on a relatively minor scale, and its variability is sufficient to exert climatological effects on Earth. The comfortable view of a steady and unchanging Sun has been replaced by a slightly confused picture of a somewhat inconstant Sun.

**1. Solar oscillation**
The distortions of the Sun's shape have been greatly exaggerated in this illustration. The oscillations consist of alternate bulges at the poles and at the equator.

**2. Changes in the Sun's diameter**
These graphs show the mean annual semi-diameter of the Sun in the horizontal (**A**) and vertical planes (**B**). The measurements made at Greenwich were obtained by measuring the time taken by the Sun to cross the meridian (for the horizontal value), and by measuring the angles from the zenith to the upper and lower

limbs (for the vertical value). The most reliable results are indicated by solid lines; the vertical divisions mark changes in the instruments or methods used.

**3. Transits of Mercury**
Timing the transits of Mercury provides a more accurate measure of the solar diameter than that used for diagram 2, but the results are still not conclusive. Readings obtained in this way are shown here as black dots, with vertical lines representing the uncertainties of each value. The open circles give values deduced from total eclipses (*see* diagram 4).

**4. Baily's beads**
An exaggerated profile of the Moon has been superimposed on a print made from a frame of cine film, recording the eclipse of 20

May 1966. The brightness of the Baily's beads corresponds with the depth of the lunar valleys. The film has been used to measure the solar diameter.

4

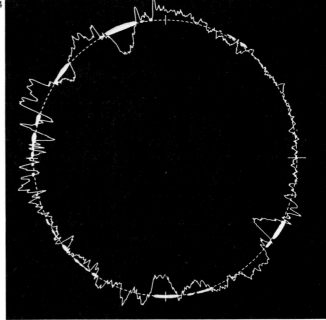

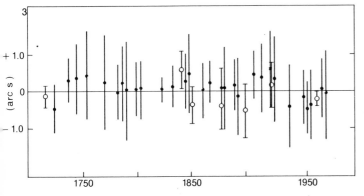

# Life Cycle of the Sun

The problem of understanding the life cycle of the Sun involves the more general problem of the evolution of stars. Since stars live for up to several thousand million years the evolution of a single star clearly cannot be observed from birth through to death. However, it is possible to examine a large cross-section of stars at different stages in their individual life cycles and to try to build up a picture of the stages through which they pass. The first step is to arrange stars into groups: the most widely used system of classification is the "MKK" system (after Morgan, Keenan and Kellman), also known as the Yerkes system. This system divides stars into 10 principal classes, or "spectral types", according to the relative strength or weakness of certain absorption-lines in their spectra (*see* diagram 1). The various spectral types are designated by a letter (O, B, A, F, G, K, M, R, N, S), corresponding to a decreasing level of effective temperature from O (hottest) to S (coolest); each class is further subdivided into 10 subclasses numbered from 0 to 9. According to this system, the Sun's spectral class is G2.

**Hertzsprung-Russell diagram**

If the spectral types (or temperatures) of a large number of stars are plotted against their luminosities (or against their absolute magnitudes) several interesting features emerge in the resulting "Hertzsprung-Russell diagram" (*see* diagram 2). Most stars lie in a narrow band sloping from the upper left of the graph to the lower right: in other words, the hotter they are the brighter they are. The Sun belongs to this group, known as the "main sequence". There is also a sizeable group of stars lying above and to the right of the main sequence, representing cool stars very much larger than main sequence stars of the same spectral class (in other words, with the same temperature), and therefore much more luminous: these are known as "red giants". Another important group lies well below and to the left of the main sequence. These are "white dwarfs"—very hot stars whose sizes are comparable with that of the Earth. Their luminosities are therefore very low.

It turns out that these three "types" of star represent different stages through which a star like the Sun will pass during the course of its life. A star is believed to form when an interstellar gas cloud collapses under its own weight. The temperature of the cloud increases as the collapse proceeds and a protostar is formed. The luminosity of the newly forming star is provided by the release of gravitational energy, but as the contraction proceeds the density at the center becomes large and the central temperature eventually reaches about $10^7$ K, at which point nuclear fusion reactions can begin, converting hydrogen to helium. These reactions are capable of supplying sufficient energy to halt the shrinkage of the star as the thermal gas pressure balances the gravitational pull. A state of equilibrium is reached and the star becomes a stable main sequence star with a temperature and luminosity determined by its initial mass: the more massive the star the hotter and brighter it will be. It remains in this state with little change of surface properties for as long as there is sufficient hydrogen fuel in the core to sustain its output. For a star like the Sun, the main sequence stage is thought to last for about $10^{10}$ years. Since the estimated age of the Sun is about $4.6 \times 10^9$ years, it should remain on the main sequence for a further 5 or $6 \times 10^9$ years.

Eventually, the core of the star will become exhausted of hydrogen and clogged with helium "ash". The core, shrinking under the weight of the overlying layers of material, will heat up hydrogen in a shell surrounding the core to a level where nuclear fusion is possible. Indeed, it appears that the rate of energy production will increase substantially, causing the star to expand until internal pressure and gravitational attraction are once again in a state of balance. This process causes the star to expand into a red giant, and its luminosity increases perhaps 1,000 times. When the Sun reaches this stage, the planet Mercury will be swallowed up into the solar globe and life on Earth will become impossible.

When the core temperature of a red giant reaches about $10^8$ K a new core reaction commences, converting carbon to helium; this process ushers in another relatively stable state of existence. A star cannot remain a red giant for long, however, since the enhanced luminosity implies that it is using up its reserves of nuclear fuel at a greatly increased rate. After at most a few hundred million years as a red giant, the shrinking dead core encompasses essentially the whole star and the star, its nuclear energy spent, collapses to

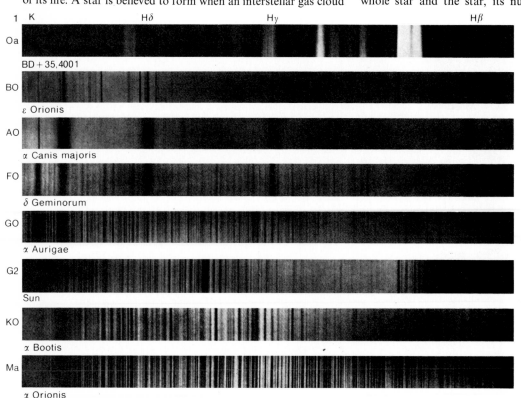

**1. Spectral classes O to M**
The peak of continuum emission shifts towards longer wavelengths as the temperature decreases from O type stars (hottest) to M type (coldest). The main absorption-line features are:
O: ionized and neutral helium, ionized metals, weak hydrogen
B: neutral helium, ionized metals, hydrogen stronger
A: hydrogen dominant, ionized metals
F: hydrogen weaker, neutral and singly ionized metals (particularly calcium)
G: singly ionized calcium prominent, hydrogen weaker, neutral metals
K: strong metallic lines, some molecular bands
M: neutral metals, strong titanium oxide bands.

**2. Hertzsprung-Russell diagram**
Stars pass through various positions on the diagram in the course of their lifetimes, spending the longest period as part of the "main sequence". The Sun's evolutionary track up to the present is shown as a bold line; its predicted development (dashed line) is shown schematically.

Diagram 1 labels: K, Hδ, Hγ, Hβ (top), with rows labeled Oa (BD + 35,4001), BO (ε Orionis), AO (α Canis majoris), FO (δ Geminorum), GO (α Aurigae), G2 (Sun), KO (α Bootis), Ma (α Orionis).

become a white dwarf. The shrinkage is halted not by thermal pressure but by the pressure of the high-speed electrons flying around in the interior as a result of the high density. When this stage is reached, the mass is not substantially less than before, but the density is greater by a factor of about a million. A teaspoonful of white dwarf material, if brought back to Earth, would weigh several tonnes. As a white dwarf, the Sun will slowly radiate away internal energy, cooling down over thousands of millions of years eventually to become a cold dark body—a black dwarf.

This is the sequence of events for the Sun sketched out by current theories of stellar evolution. However, the questions that are now being raised about the mechanisms operating in the solar interior may yet prove to have far-reaching consequences for the whole question of stellar evolution.

## Origin of the solar system
The account of stellar evolution that has emerged involves a number of problems. Clearly nature has solved the problems, for stars exist; and there is good observational evidence for new stars being located within clouds of interstellar material, but it is difficult to see how stars can form from gas clouds: unless the initial cloud is tens or even hundreds of thousands of solar masses, the internal pressure due to temperature, magnetic fields and rotation would be too great for it to collapse under its own weight. Some kind of trigger would appear to be required to cause clouds to collapse.

One possible trigger involves the spiral arms of the galaxy, in which regions of enhanced density, comprising gas and stars, are separated by less dense regions between the arms. As a cloud of gas rotates about the galactic center it passes in and out of these spiral arms. When it collides with a denser region the cloud will be compressed, and this may trigger its collapse. The collapse is thought to lead to fragmentation of the cloud, and to the formation of individual massive stars. Such stars go through their life cycles very rapidly indeed (about 10 million years) and many are believed to blow themselves violently apart in cataclysmic eruptions known as "supernovae". A supernova scatters into the interstellar regions the heavy elements built up in its interior by nuclear reactions,

contaminating the original material of the galaxy (believed to be hydrogen and helium). Later generations of stars such as the Sun are constructed out of this contaminated mix. The supernova eruption, sending shock waves through the surrounding interstellar medium, provides one of a number of possible mechanisms for compressing smaller clouds and leading to the formation of individual stars of solar mass.

Useful evidence is provided by meteorites, presumed to be the oldest solid material in the Solar System. They contain the decay products of short-lived radioactive materials, and the relative abundance of these elements suggests that a supernova occurred within a million years of the solidification of the meteorite materials. This supernova may well have been the trigger that led to the formation of the Sun and the rest of the Solar System.

Although there is considerable dispute as to the mode of formation of the Solar System, the concensus view is that as the proto-solar cloud collapsed it formed a spinning flattened nebula with a hot center which eventually became the Sun. The angular momentum of the Sun was transferred to the surrounding disc of gas by the action of the solar magnetic field and by friction, slowing the Sun to its present modest rate of revolution and accelerating the surrounding material. The Sun then contracted to become a main sequence star, adjusting its diameter to suit the rate of energy generation and decreasing in luminosity as it did so. As the outer cloud cooled, material began to solidify and eventually accreted to form the planets; the outer giant planets, moving in the cooler regions of the solar system, retained vast envelopes of hydrogen and helium which the inner planets, having higher temperatures and lower gravities, were unable to do. Young stars are known to have very strong stellar winds, and the strong solar wind of this early phase is believed to have swept away most of the residual gas from the system, leaving it essentially as it is today. According to this view the Sun and planets originated at the same time and from the same initial cloud of interstellar material. There are, however, many uncertain links in the chain and no-one can be confident that the mystery has been solved. Revolutionary changes of opinion may lie only just around the corner.

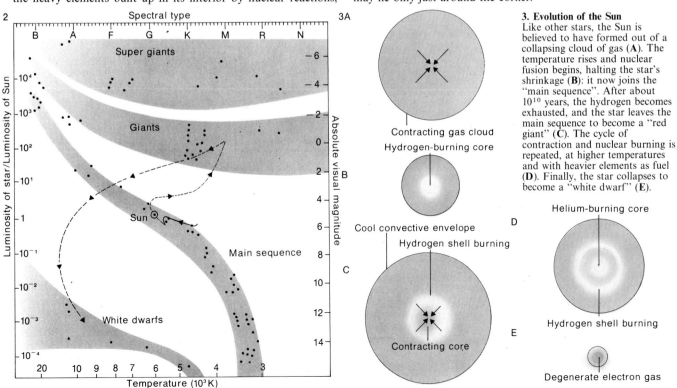

### 3. Evolution of the Sun
Like other stars, the Sun is believed to have formed out of a collapsing cloud of gas (**A**). The temperature rises and nuclear fusion begins, halting the star's shrinkage (**B**): it now joins the "main sequence". After about $10^{10}$ years, the hydrogen becomes exhausted, and the star leaves the main sequence to become a "red giant" (**C**). The cycle of contraction and nuclear burning is repeated, at higher temperatures and with heavier elements as fuel (**D**). Finally, the star collapses to become a "white dwarf" (**E**).

# Glossary

**Albedo** The ratio of the amount of light reflected by a body to the amount of light incident on it; a measure of the reflecting power of a body. A perfect reflector would have an albedo of 1. The albedos of the planets are as follows: Mercury 0.06, Venus 0.76, Earth 0.39, Mars 0.16, Jupiter 0.43, Saturn 0.61, Uranus 0.35, Neptune 0.35, Pluto 0.5.

**Allotropy** The property in a chemical element of existing in different forms, with distinct physical properties but capable of forming identical chemical compounds. Ozone, for example, is an allotropic form of oxygen.

**Altitude** In astronomy, the angular distance of a celestial body from the horizon. In conjunction with a measurement of AZIMUTH, it describes the position of an object in the sky at a given moment.

**Aphelion** The point or moment of greatest distance from the Sun of an orbiting body such as a planet. The opposite of PERIHELION.

**Asteroid** One of a large number of rocky objects, smaller than a planet but larger than a METEORITE, in orbit around the Sun. Also known as "minor planets", over 99 percent of the asteroids in the SOLAR SYSTEM lie in a belt situated between the orbits of Mars and Jupiter.

**Astronomical unit** A unit of distance defined by the mean distance of the Earth from the Sun and equal to 149,597,870 km.

**Azimuth** The angular distance along the horizon, measured in an eastward direction, between a point due north and the point at which a vertical line through a celestial object meets the horizon. (This is the normal convention for an observer in the northern hemisphere; other conventions are sometimes followed.) *See also* ALTITUDE.

**Barycenter** The center of gravity of a system of massive bodies; the barycenter of the Earth–Moon system, for example, lies at a point within the Earth's globe.

**Black body** An idealized body which reflects none of the radiation falling on it. Such a body would be a perfect absorber of radiation, and would emit a SPECTRUM determined solely by its temperature.

**Bode's Law** A curious numerical relationship between the distances of the various planets from the Sun. The law is often expressed in the form:
$$r_n = 0.4 + 0.3 \times 2^n,$$
where $r_n$ is the distance of the planet from the Sun and n is

$-\infty$ 0, 1, 2, 3 . . . in turn. The resulting values correspond surprisingly closely with the actual distances, but most astronomers consider this to be merely a coincidence.

**Celestial equator** The circle formed by the projection of the Earth's equator onto the surface of the CELESTIAL SPHERE.

**Celestial sphere** An imaginary sphere, centered on the Earth, onto whose surface the stars may be considered, for the purposes of positional measurement and calculation, to be fixed.

**Chromosphere** The layer of the Sun's atmosphere lying above the PHOTOSPHERE and below the CORONA.

**Comet** A type of heavenly body in orbit around the Sun, with several characteristics that distinguish it from the planets, satellites or asteroids. Comets typically have highly eccentric orbits, and some of them become bright objects in the sky as they approach PERIHELION, sometimes with a distinctive "tail". They are made up of a "nucleus" with a surrounding cloud of dust and gas which forms the "coma".

**Conjunction** The near or exact alignment of two astronomical bodies in the sky. Also used to describe an alignment between a planet and the Sun as seen from Earth. When the planet passes behind the Sun, the conjunction is called "superior"; in the special case of Mercury or Venus passing between the Sun and the Earth, the conjunction is called "inferior".

**Coriolis effect** The apparent deflection of a body moving in a rotating coordinate system. For example, a projectile fired northward from the Earth's equator will appear to be deflected to the east, because the point on the equator from which it is fired will be rotating faster. than its target to the north. The Coriolis effect plays an important part in determining the directions of wind and ocean currents.

**Corona** The outermost part of the Sun's atmosphere. It is visible to the naked eye only during a total eclipse of the Sun, when it has the appearance of a halo around the Sun's obscured disc. The corona is the source of the SOLAR WIND.

**Cosmic rays** Extremely energetic atomic particles, principally protons, travelling through space at speeds approaching the speed of light. A proportion of cosmic rays come from the Sun, while the rest originate somewhere outside the SOLAR SYSTEM, possibly in violent events in the GALAXY.

**Culmination** The maximum altitude of a celestial body above the horizon.

**Declination** The angular distance of a celestial body from the CELESTIAL EQUATOR; one of the two celestial coordinates, roughly equivalent to latitude on the Earth, used to represent the position of a celestial object. *See also* RIGHT ASCENSION.

**Doppler effect** The apparent shift in the frequency of waves that occurs when there is relative motion between the source and the observer. A receding source will appear to emit waves of longer wavelength (or lower frequency) than it would if it were stationary; with an approaching source the effect is reversed, and the wavelength appears to be shorter (higher frequency).

**Eclipse** The partial or total disappearance of a celestial body either behind a nonluminous body or into its shadow. A solar eclipse, for example, occurs when the Sun is obscured by the Moon's disc, while a lunar eclipse takes place when the Moon passes through the cone of shadow cast by the Earth.

**Ecliptic** The circle on the CELESTIAL SPHERE defined by the Sun's apparent annual motion against the stellar background. The ecliptic represents the plane in which the Earth orbits the Sun and, because the Earth's rotational axis is tilted, the ecliptic is inclined to the celestial equator at an angle, known as the "obliquity of the ecliptic", which is equal to about $23\frac{1}{2}°$.

**Electromagnetic radiation** Radiation in the form of waves associated with electric and magnetic disturbances, which may be manifested in a variety of forms, such as light, X-rays and radio waves, depending on the wavelength. The electric and magnetic components are often represented as two waves oscillating in different planes at right angles to one another.

**Elongation** The angular distance of a planet from the Sun, or of a satellite from its primary planet.

**Equation of time** The difference between the apparent solar time and the mean time; the value of the equation of time varies throughout the year from about $-14\frac{1}{4}$ min to about $+16\frac{1}{4}$ min.

**Exosphere** The outermost region of the Earth's atmosphere, beyond the IONOSPHERE.

**Faculae** Bright patches on the PHOTOSPHERE of the Sun, normally associated with SUNSPOT groups.

**First Point of Aries** *See* VERNAL EQUINOX.

**Flares** Sudden brilliant outbursts in the outer part of the Sun's atmosphere, typically lasting only a few minutes. Generally associated with SUNSPOTS, they give rise to a type of COSMIC RAYS.

**Fraunhofer lines** Dark lines appearing in the spectrum of the Sun, resulting from the absorption of certain wavelengths of light by elements in the outer parts of the Sun's atmosphere.

**Galaxy** A large system of stars. The term "The Galaxy" refers to the particular galaxy of which the Sun is a member.

**Hertzsprung-Russell diagram** A graph on which is plotted the LUMINOSITY of stars against their temperature or spectral type. The diagram reveals that for a given spectral type, temperature is not randomly distributed. For the most numerous group of stars (the so-called "main sequence" stars) the higher the temperature, the brighter, in general, is the star; other groupings in the H-R diagram represent stellar types such as Red Giants and White Dwarfs which do not obey this general rule.

**Ion** An atom that is electrically charged as a result of having lost or gained one or more electrons.

**Ionosphere** The region of the Earth's atmosphere, extending from approximately 80 km to 500 km above the surface, in which radiation from the Sun ionizes a substantial proportion of air molecules. *See also* ION.

**Librations** Apparent oscillations of the Moon as a result of which an Earth-based observer can see the surface from a slightly different angle at different times. Over a period of time, a total of about 59 percent of the Moon's surface can be seen from Earth.

**Light-year** A unit of distance defined by the distance travelled by light *in vacuo* in a year, equal to $9.4607 \times 10^{12}$ km or 63,240 ASTRONOMICAL UNITS. In astronomy the more commonly used unit for large distances is the PARSEC, which is equal to 3.2616 light-years.

**Limb** The edge of the visible disc of a celestial body.

**Luminosity** The total amount of energy emitted by a star per unit of time.

**Magnetosphere** The region around a planet within which its magnetic field predominates over the magnetic field of the surrounding interplanetary region.

**Magnitude** A measure of the brightness of a star or other celestial body on a numerical scale which decreases as the brightness increases. The faintest stars visible to the naked eye on a clear night are of magnitude 6; the brightest have a mean magnitude of 1. The "absolute" magnitude of a star is defined as the apparent magnitude it would have if viewed from a standard distance of 10 PARSECS.

**Meridian** A great circle passing through the poles either of the Earth or of the CELESTIAL SPHERE. In astronomical usage, the term usually refers to the "observer's meridian", which passes through the observer's ZENITH.

**Meteor** A small particle of interplanetary material that leaves a bright trail across the sky as it burns up on entering the Earth's atmosphere.

**Meteorite** The remains of a METEOR that reaches the surface.

**Meteoroid** A small lump of solid meteoritic material in space.

**Nodes** The points at which two great circles on the CELESTIAL SPHERE intersect; in particular, the points at which the orbit of a body, such as a planet or the Moon, crosses the ECLIPTIC.

**Occultation** The temporary disappearance of one celestial body, usually a star, behind another, usually a planet or moon. A solar eclipse is a particular case of an occultation.

**Opposition** The position of a planet in its orbit when the Earth lies on a direct line between the planet and the Sun. A planet is best placed for observation when it is at opposition.

**Parallax** The apparent change in the position of an object due to an actual change in the position of the observer. Measurement of parallax allows the distances of distant objects to be determined.

**Parsec** A large unit of distance defined as the distance at which a star would have an annual PARALLAX of one second of arc, and equal to $3.0857 \times 10^{13}$ km, 206,265 ASTRONOMICAL UNITS, or 3.2616 LIGHT-YEARS.

**Penumbra** The region of partial shadow that is formed around the region of total shadow when the source of illumination is of finite size. *See also* UMBRA. The term is also used to describe the outer part of SUNSPOTS.

**Perihelion** The point or moment of closest approach to the Sun of an orbiting body such as a planet. The opposite of APHELION.

**Perturbations** Irregularities in the orbital motion of a body due to the gravitational influence of other orbiting bodies.

**Phase angle** The angle defined by the position of the Sun, a body, and the Earth, measured at the body.

**Photosphere** The intensely luminous layer of the Sun that forms its visible surface.

**Plage** Bright areas of the PHOTOSPHERE associated with active areas on the Sun, caused by the presence of gas considerably hotter than its surroundings.

**Planet** One of the nine medium-sized bodies (including the Earth) which orbit the Sun; a similar body orbiting any other star. Unlike stars, planets do not emit their own heat or light from thermonuclear reactions in their interiors. The word "planet" is derived from a Greek word meaning "wanderer": the planets are seen to move against the background of fixed stars. An "inferior" planet is one whose orbit lies within that of the Earth, while a "superior" planet moves outside the Earth's orbit.

**Polarization** A special condition of ELECTROMAGNETIC RADIATION. Radiation (such as light) may be resolved into two components, one electrical, the other magnetic, at right angles to one another. When the radiation is unpolarized, the components vibrate in every direction, but if the radiation is "plane polarized", all the electrical components are arranged in planes parallel to each other, with their associated magnetic components lying at right angles to them. Other types of polarization, such as "circular" and "elliptical", are also possible.

**Quadrature** The position of the Moon or an outer planet when its ELONGATION is 90°.

**Right ascension** (R.A.) The angle, measured eastward along the CELESTIAL EQUATOR in units of hours, minutes and seconds, between the VERNAL EQUINOX and the point at which the MERIDIAN through a celestial object intersects the celestial equator. Right ascension is roughly equivalent to longitude on the Earth, and in conjunction with one other coordinate, DECLINATION, specifies the exact position of an object in the sky.

**Roche limit** The critical distance from the center of a planet within which gravitational forces would be insufficient to prevent a satellite from being broken up by tidal forces. For a satellite with the same density as the parent planet, the Roche limit lies at 2.4

times the radius of the planet.

**Saros** An interval of 6,583 days (equal to 18 years 11.3 days) after which the Sun, the Moon and the Earth return almost exactly to their previous relative positions. Consequently, the Saros period marks the interval between successive ECLIPSES of similar type and circumstance.

**Sidereal period** The time taken for a body to complete one orbit, as measured against the background of fixed stars. *See also* SYNODIC PERIOD.

**Sidereal time** A system of measurement of time based on the Earth's period of rotation, measured against the background of fixed stars. The sidereal day is taken to begin at the moment at which the VERNAL EQUINOX crosses the observer's MERIDIAN.

**Solar constant** The amount of energy per second that would be received in the form of solar radiation over one square meter of the Earth's surface at the Earth's mean distance from the Sun, if no radiation was absorbed by the atmosphere.

**Solar cycle** The periodic variation of solar activity, as manifested in the number of SUNSPOTS, the frequency of solar FLARES and various other solar phenomena. The cycle has an average period of about 11 years.

**Solar System** The system made up of the Sun, the planets (Mercury, Venus, Earth, Mars, Jupiter, Saturn, Uranus, Neptune and Pluto) together with their satellites, the ASTEROIDS, COMETS, METEOROIDS and interplanetary material.

**Solar wind** An electrically charged stream of atomic particles, mainly protons and electrons, emitted by the Sun.

**Solstices** The two points on the ecliptic of maximum or minimum DECLINATION; the times at which the Sun reaches these points along its annual path. The summer solstice (corresponding to the maximum declination) falls around 21 June, the winter solstice (minimum declination) around 21 December.

**Spectrum** The range of wavelengths or frequencies present in a sample of ELECTROMAGNETIC RADIATION. Visible radiation (i.e. light) may be resolved into its component wavelengths by passing it through a prism; white light will be spread out into a band of colors. A glowing gas under low pressure will emit radiation only at certain specific wavelengths, which appear as bright, isolated "emission" lines in

its spectrum; similarly, it will only absorb radiation at these same wavelengths. When radiation is absorbed from a "continuous" spectrum, black "absorption" lines appear. *See also* FRAUNHOFER LINES.

**Stratosphere** The region of the Earth's atmosphere, extending from about 15 km to 50 km above the Earth's surface, between the TROPOSPHERE and the mesosphere.

**Sunspots** Large transient patches on the PHOTOSPHERE of the Sun which appear black in contrast with the surrounding regions. The number of sunspots varies in a periodic way (*see* SOLAR CYCLE).

**Synchrotron radiation** Radiation emitted by electrons travelling in a strong magnetic field at speeds approaching the speed of light.

**Synodic period** The interval between successive CONJUNCTIONS or, more generally, between similar configurations of a celestial body, the Sun and the Earth.

**Tektites** Small glassy objects, found in a few restricted areas of the Earth, whose origin remains a mystery; believed to be associated with METEORITE impacts on Earth.

**Terminator** The boundary between the dark and the sunlit hemispheres of a planet or satellite.

**Tropopause** The boundary between the TROPOSPHERE and the STRATOSPHERE.

**Troposphere** The lowest layer of the Earth's atmosphere, within which temperature decreases with increasing altitude. It extends to a height of about 15 km.

**Umbra** The dark central region of a shadow. *See also* PENUMBRA.

**Vernal equinox** The point on the CELESTIAL SPHERE at which the ECLIPTIC crosses the CELESTIAL EQUATOR from south to north (where the direction is defined by the Sun's motion). Also known as the First Point of Aries.

**Zeeman effect** The splitting of spectral lines (*see* SPECTRUM) when emission or absorption occurs in the presence of a strong magnetic field.

**Zenith** The point on the CELESTIAL SPHERE directly above the observer.

**Zodiac** A belt on the CELESTIAL SPHERE extending by about 8° on either side of the ECLIPTIC, marking the region within which the Sun and the planets are always to be found. The zodiac is divided into 12 equal zones which are named after 12 constellations.

# Observing the Sun

In many ways the Sun is an ideal object for the amateur observer to study. Weather permitting, it is available all day every day, and large telescopes are not required for its observation. A 50 mm aperture telescope will suffice to show sunspots, and apertures in the range of 75 mm to 150 mm are the most useful size for the amateur, refractors being rather more convenient for this purpose than reflectors.

**It cannot be stressed too strongly that it is dangerous to look at the Sun—even without a telescope—and on no account should an observer risk looking at the Sun through a telescope, even for the briefest instant. Serious and irreparable eye damage, and quite probably permanent blindness, would be the result.**

Dark filters placed over the eyepiece are not safe; if they should crack, there is little chance of the observer removing his eye in time to avoid damage. The Herschel wedge, or solar diagonal, contains an unsilvered glass wedge that passes most of the light and heat straight through and reflects only a small proportion to an eyepiece; even so, a dark filter is required at the eyepiece. Thin sheets of polyester-type material coated with aluminium may be purchased commercially. Placed over the full aperture of the telescope, these are not subject to concentrated heating, but great care has to be taken to ensure that the film is fully effective over its full width, and that it cannot be ruptured or blown away.

Without doubt the safest way to observe the Sun is by the *projection method*, the focus of the telescope being adjusted to produce a sharp image on a white card or other form of screen held beyond the eyepiece. A card placed round the body of the telescope is necessary to cast a shadow on to the screen so that the projected image is not obscured by direct sunlight; many observers construct a projection box to attach to the rear of the telescope to provide a shaded compartment into which to project the image.

One straightforward type of observation is to make daily records of sunspot activity. The projected image should be drawn to a standard size (a 150 mm diameter disc is most convenient), the positions of sunspots, groups, and faculae being marked in first and details added later, possibly with the aid of a higher magnification. If sunspots are allowed to drift across the field of view, their direction of motion can be used to establish the east–west direction so that the observer may orientate his drawing to allow for the tilt of the solar disc at different times of the year (*see pages 38–39*).

The Sun may be photographed at prime focus or by eyepiece projection with the camera body attached to the telescope, but care must be taken to avoid overheating the camera itself. Alternatively, the image projected onto a screen may be photographed.

Commercially available $H_\alpha$ filters have opened up a fruitful field for the amateur in the plotting of plages, filaments, prominences and flares. Solar radio astronomy also offers great scope for the amateur. Solar bursts and noise storms lie within the capacity of fairly basic equipment—strong bursts may be picked up even with a simple yagi aerial and a television receiver. Most amateurs use equipment operating at frequencies of a few tens to a few hundreds of MHz, depending on the equipment and type of aerial available.

It is not unknown for amateurs to detect significant events that have escaped the monitoring activities of professional observatories, and there are good links between amateur solar radio observers and professional institutions. As with all forms of observing, the results obtained are of much greater value if observers collaborate to obtain maximum coverage of events. In the United Kingdom the British Astronomical Association and the Junior Astronomical Society maintain solar sections with a director who correlates observations sent in by many observers. The Radio Astronomy section of the BAA coordinates solar radio observations. In the United States the American Association of Variable Star Observers fulfils a similar function. Equivalent organizations exist in other nations.

# Bibliography

## Books

Abetti, G., *Solar Research* (Eyre & Spottiswoode, 1962)

Abetti, G., *The Sun* (Faber & Faber, 1955)

Allen, C.W., *Astrophysical Quantities* (3rd edition, Athlone Press, 1973)

Baxter, W.M., *The Sun and the Amateur Astronomer* (Lutterworth Press, 1963)

Brandt, J.C., *Introduction to the Solar Wind* (W.H. Freeman, 1970)

Bray, R.J., and Loughhead, R.E., *The Solar Chromosphere* (Chapman and Hall, 1974)

Bray, R.J., and Loughhead, R.E., *Sunspots* (Dover, 1979)

Bruzek, A. and Durant, C.J., eds, *Illustrated Glossary for Solar and Solar-terrestrial Physics* (D. Reidel, 1977)

Bumba, V., and Kleczek, J., *Basic Mechanisms of Solar Activity* (D. Reidel, 1975)

Eddy, J.A., *A New Sun—the Solar Results from Skylab* (NASA-SP-402, 1979)

Eddy, J.A., ed., *The New Solar Physics* (American Association for the Advancement of Science, published by Westview Press, 1978)

Gibson, E.G., *The Quiet Sun* (NASA-SP-303, 1973)

Hargreaves, J.K., *The Upper Atmosphere and Solar-terrestrial Relations* (Van Nostrand Reinhold, 1979)

Hundhausen, A.J., *Coronal Expansion and the Solar Wind* (Springer-Verlag, 1972)

Kane, S.R., *Solar Gamma, X-, and EUV radiation* (D. Reidel, 1975)

Kruger, A., *Introduction to Solar Radio Astronomy and Radio Physics* (D. Reidel, 1979)

Kundu, M.R., *Solar Radio Astronomy* (Interscience, 1965)

Meadows, A.J., *Early Solar Physics* (Pergamon Press, 1970)

Menzel, D.H., *Our Sun* (Harvard University Press, 1959)

Meuss, J., *et al.*, *Canon of Solar Eclipses* (Pergamon, 1966)

Newton, H.W., *The Face of the Sun* (Penguin Books, 1958)

Ramaty, R. and Stone, R.G., eds., *High Energy Phenomena on the Sun* (NASA-SP-342, 1973)

Smith, A.G., *Radio Exploration of the Sun* (Van Nostrand Reinhold, 1967)

Sonett, C.P., *et al.*, eds., *Solar Wind* (NASA-SP-308, 1972)

Sturrock, P.A., ed., *Solar Flares*, Proceedings of the Second Skylab Workshop (University of Colorado, 1980)

Svestka, Z., *Solar Flares* (D. Reidel, 1976)

White, O.R., ed., *The Solar Output and its Variation* (Colorado Associated University Press, 1977)

Zirin, H., *The Solar Atmosphere* (Blaisdell Publishing Company, 1966)

*The Handbook of the British Astronomical Association* (1980)

## Articles

Albregtsen, F., and Maltby, P., "New Light on Sunspot Darkness and the Solar Cycle", *Nature*, **274**, 41–42 (1978)

Allen, J.A. Van, "Interplanetary Particles and Fields", *Scientific American*, **233**, 160–173 (1975)

Bahcall, J.N. & Davis Jr., R., "Solar Neutrinos—a scientific Puzzle", *Science*, **191**, 264–267 (1976)

Belitsky, B., "Soviet neutrino astronomy", *Spaceflight*, **19**, 311–312 (1977)

Brookes, J.R., *et al.*, "Observations of free oscillations of the Sun", *Nature*, **259**, 92–95 (1976)

Brown, J.C., and Smith, D.F., "Solar Flares", *Reports on Progress in Physics*, **43**, 125–197 (1980)

Christensen-Dalsgaard, J., and Gough, D.O., "Towards a heliological inverse problem", *Nature*, **259**, 89–92 (1976)

Chung-Cheih Cheng, "Spacial distribution of XUV emission and density in a loop prominence", *Solar Physics*, **65**, 347–356 (1980)

Clark, D., "Our inconstant Sun", *New Scientist*, **81**, 168–170 (1979)

Davis Jr., R., and Evans Jr., J.C., "Neutrinos from the Sun", in *The

*New Solar Physics*, Eddy, J.A., ed. (Westview, 1978)

Dennis, B.R., *et al.*, "The Solar Flare of 1980, March 29 at 0918 UT as observed with the hard X-ray burst spectrometer on the Solar Maximum Mission", *Astrophys. J., 244*, (Letts), L167–170 (1981)

Dicke, R.H., "The Clock Inside the Sun", *New Scientist*, **83**, 12–14 (1979)

Dunham, D.W., *et al.*, "Observations of a Probable change in the solar radius between 1715 and 1979", *Science*, **210**, 1243–1245 (1980)

Eddy, J.A., "The Case of the Missing Sunspots", *Scientific American*, **236**, 80–92 (1977)

Eddy, J.A., "Historical and Arboreal Evidence for a Changing Sun", in *The New Solar Physics*, Eddy, J.A., ed. (Westview 1978)

Eddy, J.A., and Boornazian, A.A., "Secular decrease in the solar diameter 1863–1953", Bull., *Am. Astron. Soc.*, **11**, 437 (1979)

Friedman, H., "Sun and Earth: A New View of Climate", *Astronautics and Aeronautics*, 58–63 (February 1979)

Gough, D., "The Shivering Sun Opens its Heart", *New Scientist*, **70** 590–592 (1976)

Gough, D., "Climate and Variability in the Solar Constant", *Nature*, **288**, 639–640, 1980

Grec, G. *et al.*, "Solar Oscillations: Full Disk Observations from the Geographic South Pole", *Nature*, **188**, 541–544 (1980)

Hartlin, B.K., "In Search of Solar Neutrinos", *Science*, **204**, 42–44 (1979)

Hill, H.A., "Seismic Sounding of the Sun", in *The New Solar Physics*, Eddy, J.A., ed. (Westview 1978)

Hood, A.W. and Priest, E.R., "Magnetic Instability or Coronal Arcades as the Origin of 2-ribbon Flares", *Solar Physics*, **66**, 113–134 (1980)

House, L.L., *et al.*, "Studies of the Corona with the Solar Maximum Mission Coronograph/Polarimeter", *Astrophys. J.*, **244** (Letts), L117–121 (1981)

Howard, R., "Large-Scale Solar Magnetic Fields", *Ann. Rev. Astron. Astrophys.*, **15**, 153–173 (1977)

Howard, R., "The Rotation of the Sun", *Scientific American*, **232**, 106–114 (1975)

Howard, R., "A Possible Variation of the Solar Rotation with the Activity Cycle", *Astrophysical Journal*, **210** L159–L161 (1976)

Hoyng, P., *et al.*, "Hard X-ray Imaging of Two Flares in Active Region 2372", *Astrophys. J.* **244** (Letts), L153–156 (1981)

Hundhausen, A.J., "Streams, Sectors and Solar Magnetism", in *The New Solar Physics*, Eddy, J.A., ed. (Westview, 1978)

Jordan, C., "The Outer Layers of the Sun", *Science Progress*, **67**, Spring 1981

Markson, R., "Solar Modulation of atmospheric electrification and possible implications for the Sun–weather relationship", *Nature*, **273**, 103–109 (1978)

Noyes, R.W., "New Developments in Solar Research", in *Frontiers of Astrophysics*, Avrett, E.H., ed. (Harvard University Press, 1976)

Parker, E.N., "The Origin of Solar Activity", *Ann. Rev. Astron. Astrophys.*, **15**, 45–68 (1977)

Parkinson, J.H., "The constancy of solar diameter over the past 250 years", *Nature*, **288**, 548–551 (1980)

Parkinson, J.H., "What's wrong with the Sun", *New Scientist*, **86**, 201–204 (1980)

Pasachoff, J.M., "The Solar Corona", *Scientific American*, **229**, 68–79 (1973)

Severny, A.B., *et al.*, "Observations of Solar Pulsations", *Nature*, **259**, 87–89 (1976)

Shapiro, I.I., "Is the Sun Shrinking?", *Science*, **208**, 51–53 (1980)

Sheeley Jr., N.R., "Temporal variations of loop structures in the solar atmosphere", *Solar Physics*, **66**, 79–87 (1980)

Siscoe, G.L., "Solar-terrestrial influences on weather and climate", *Nature*, **276**, 348–352 (1978)

Spicer, D.S., "An Unstable Arch Model of a Solar Flare", *Solar Physics*, **53**, 305 (1977)

Spiegel, E.A., and Weiss, N.O., "Magnetic Activity and Variations in Solar Luminosity", *Nature*, **287**, 616–617 (1980)

Stuiver, M., "Solar Variability and Climatic Change During the Current Millenium", *Nature*, **286**, 868–871 (1980)

Svalgaard, L., and Wilcox, J.M., "A view of Solar Magnetic Fields, the Solar Corona, and the Solar Wind in three dimensions", *Ann. Rev. Astron. Astrophys.*, **16**, 393–428 (1978)

Svalgaard, L., and Wilcox, J.M., "Structure of the extended solar magnetic field and sunspot cycle variations in cosmic ray intensity", *Nature*, **262**, 766–768 (1976)

Tandberg-Hanssen, E., *et al.*, "Ultraviolet spectroscope and polarimetry on the Solar Maximum Mission", *Astrophys. J.* **244** (Letts), L127–132 (1981)

Vaiana, G.S., *et al.*, "X-ray observations of characteristic structures and time variations from the solar corona: preliminary results from Skylab", *Astrophys. J.* **185** (Letts), L47–L51 (1973)

Vaiana, G.S. and Rosner, R., "Recent Advances in Coronal Physics", *Ann. Rev. Astron. Astrophys.*, **16**, 393–428 (1978)

Van Beek, H.F., *et al.*, "The limb flare of 30 April, 1980 as seen by the hard X-ray imaging spectrometer", *Astrophys. J.* **244** (Letts), L157–162 (1981)

van Tend, W., "The importance of photospheric magnetic field complexity for coronal energy storage", *Solar Physics*, **66**, 21–28 (1980)

van Tend, W., "Coronal heating by prominence turbulence", *Solar Physics*, **66**, 29–37 (1980)

Wagner, W.J., *et al.*, "Radio and visible light observations of matter ejected from the Sun", *Astrophys. J.* **244** (Letts), L123–126 (1981)

Willson, R.C., and Hudson, H.S., "Variations of Solar Irradiance", *Astrophys. J.* **244** (Letts), L185–189 (1981)

Willson, R.C. *et al.*, "Observations of Solar Irradiance Variability", *Science*, **211**, 700–702 (1981)

Wilson, O.C., "The Activity Cycles of Stars", *Scientific American*, **244**, 82–91 (1981)

Withbroe, G.L. and Noyes, R.W., "Mass and Energy Flow in the Solar Chromosphere and Corona", *Ann. Rev. Astron. Astrophys.*, **15**, 363–387 (1977)

Worden, S., and Simon, G., "On the Origin of $2^h\,40^m$ solar oscillations", *Astrophys. J.* **210** (Letts), L163–166 (1976)

Zirker, J.B., "Total eclipses of the Sun", *Science*, **210**, 1313–1319 (1980)

# Solar Eclipses  1981–2000

| Date | Type of eclipse | Maximum duration | Maximum phase (if partial) | Track Begins | | Ends | | Country |
|---|---|---|---|---|---|---|---|---|
| 4 Feb 1981 | Annular | 1 min | – | 39° S | 132° E | 16° S | 78° W | South of Australia and New Zealand, Pacific Ocean |
| 31 July 1981 | Total | 2 min 3 sec | – | 42° N | 40° E | 25° N | 159° W | USSR, North Pacific Ocean |
| 25 Jan 1982 | Partial | – | 0.57 | – | | – | | Antarctic |
| 21 June 1982 | Partial | – | 0.62 | – | | – | | Antarctic |
| 20 July 1982 | Partial | – | 0.46 | – | | – | | Arctic |
| 15 Dec 1982 | Partial | – | 0.74 | – | | – | | Arctic |
| 11 June 1983 | Total | 5 min 11 sec | – | 36° S | 60° E | 18° S | 168° E | Indian Ocean, East Indies, Pacific Ocean |
| 4 Dec 1983 | Annular | 4 min | – | 34° N | 58° W | 10° N | 50° E | Atlantic Ocean, Equatorial Africa, Somalia |
| 30 May 1984 | Annular | 1 min | – | 1° N | 136° W | 28° N | 3° E | Pacific Ocean, Mexico, USA, Atlantic Ocean, Algeria |
| 22–23 Nov 1984 | Total | 1 min 59 sec | – | 0° | 128° E | 33° S | 88° W | East Indies, South Pacific Ocean |
| 19 May 1985 | Partial | – | 0.84 | – | | – | | Arctic |
| 12 Nov 1985 | Total | 1 min 55 sec | – | 52° S | 146° W | 70° S | 164° W | South Pacific Ocean, Antarctic |
| 9 Apr 1986 | Partial | – | 0.82 | – | | – | | Antarctic |
| 3 Oct 1986 | Annular/Total | 0 min 1 sec | – | 66° N | 26° W | 56° N | 28° W | North Atlantic Ocean |
| 29 Mar 1987 | Annular/Total | 0 min 56 sec | – | 47° S | 71° W | 11° N | 54° E | Argentina, Atlantic Ocean, Congo, Indian Ocean. (Total phase over part of Atlantic Ocean.) |
| 23 Sept 1987 | Annular | 2 min | – | 46° N | 68° E | 13° S | 167° W | USSR, China, Pacific Ocean |
| 18 Mar 1988 | Total | 3 min 46 sec | – | 4° S | 86° E | 54° N | 143° W | Indian Ocean, East Indies, Pacific Ocean |
| 11 Sept 1988 | Annular | 7 min | – | 1° N | 44° E | 56° S | 165° E | Indian Ocean, south of Australia, Antarctic |
| 7 Mar 1989 | Partial | – | 0.83 | – | | – | | Arctic |
| 31 Aug 1989 | Partial | – | 0.63 | – | | – | | Antarctic |
| 26 Jan 1990 | Annular | – | – | 71° S | 74° E | 48° S | 7° W | Antarctic |
| 22 July 1990 | Total | 2 min 33 sec | – | 60° N | 24° E | 30° N | 139° W | Finland, USSR, Pacific Ocean |
| 15–16 Jan 1991 | Annular | 9 min | – | 30° S | 109° E | 0° | 114° W | Australia, New Zealand, Pacific Ocean |
| 11 July 1991 | Total | 6 min 54 sec | – | 13° N | 175° W | 13° S | 46° W | Pacific Ocean, Central America, Brazil |
| 4–5 Jan 1992 | Annular | 12 min | – | 11° N | 137° E | 33° N | 118° W | Central Pacific Ocean |
| 30 June 1992 | Total | 5 min 20 sec | – | 35° S | 35° E | 51° S | 39° E | South Atlantic Ocean |
| 24 Dec 1992 | Partial | – | 0.84 | – | | – | | Arctic |
| 21 May 1993 | Partial | – | 0.74 | – | | – | | Arctic |
| 13 Nov 1993 | Partial | – | 0.93 | – | | – | | Antarctic |
| 10 May 1994 | Annular | 7 min | – | 14° N | 146° W | 32° N | 4° W | Pacific Ocean, Mexico, USA, Canada, Atlantic Ocean |
| 3 Nov 1994 | Total | 4 min 23 sec | – | 8° S | 97° W | 32° S | 47° E | Peru, Brazil, South Atlantic Ocean |
| 29 Apr 1995 | Annular | 7 min | – | 31° S | 137° W | 6° S | 23° W | South Pacific Ocean, Peru, Brazil, South Atlantic Ocean |
| 24 Oct 1995 | Total | 2 min 5 sec | – | 34° N | 51° E | 5° N | 172° E | Iran, India, East Indies, Pacific Ocean |
| 17 Apr 1996 | Partial | – | 0.88 | – | | – | | Antarctic |
| 12 Oct 1996 | Partial | – | 0.76 | – | | – | | Arctic |
| 9 Mar 1997 | Total | 2 min 50 sec | – | 49° N | 87° E | 83° N | 159° W | USSR, Arctic Ocean |
| 2 Sept 1997 | Partial | – | 0.90 | – | | – | | Antarctic |
| 26 Feb 1998 | Total | 3 min 56 sec | – | 2° S | 144° W | 30° N | 19° W | Pacific Ocean, just south of Panama, Atlantic Ocean |
| 22 Aug 1998 | Annular | 3 min | – | 1° S | 87° E | 29° S | 155° W | Indian Ocean, East Indies, Pacific Ocean |
| 16 Feb 1999 | Annular | 1 min | – | 41° S | 8° E | 13° S | 154° E | Indian Ocean, Australia, Pacific Ocean |
| 11 Aug 1999 | Total | 2 min 23 sec | – | 41° N | 65° E | 17° N | 87° E | Atlantic Ocean, England, France, Central Europe, Turkey, India |
| 5 Feb 2000 | Partial | – | – | – | | – | | Antarctic |
| 1 July 2000 | Partial | – | – | – | | – | | Antarctic |
| 31 July 2000 | Partial | – | – | – | | – | | Arctic |
| 25 Dec 2000 | Partial | – | – | – | | – | | Arctic |

# Index

Figures in Roman type refer to text entries; figures in *italic* refer to illustrations or captions.

## A

Absolute magnitude 7
Absorption lines 26–29, *50*, *65*
  chromospheric 34
  discovery of 19
  "forbidden lines" 72, 73
  Fraunhofer lines *19*, *20*, 34, 46, 72, 73
  telluric 26
  *See also* Balmer series, Brackett series, Lyman series, Paschen series
Active prominences 48, *48*, *54*
Active regions 44–45, *44*, *54*, *56*, *59*
  and flares 66
  and corona 70, 71
Airy, Sir George 17
Albedo *9*
  changes in 81
*Almagest* 16
Alpha particles 76
Alpine glaciers 81
Anaxagoras 16, *16*, 17
Annular eclipse *See* Eclipse
Antapex 6
Antares 7
Anticyclones 80
Apex 6
Aphelion 8
Apogee 12, *12*, *13*
Apollo 24
Apollo Belvedere *5*
Apparent magnitude 7
Arch prominence *47*, 48
Argon 83
Aristarchus 16–17, *16*
Asteroids 8, *9*, 17
*Astronomia Nova* 17
Astronomical unit 8, 17
Atmosphere
  of the Earth 78–81
    and solar flares 29, 68
    and solar wind 76
    hampers observations 22, *22*
    vorticity 80
  of the Sun *See* Photosphere, Chromosphere, etc.
Atomic structure
  and spectral features 26–27, *26*
  and solar energy 82–83
Aurora *80*
  and flares *66*
  and solar wind 76
  causes of 80

## B

Babcock, H. W. & H. D. 20, 42
  *See also* Leighton-Babcock model
Babylonians 16
Baily's beads 50, 65
  and solar diameter 85, *85*
Balloons 22
Balmer series 27, *27*
Bartels, J. 76
Barycenter 8

Berkowski 19
Beryllium 82, *82*
Big Bear Solar Observatory 22, *44*
Bipolar magnetic regions (BMRs) 44–45
Bipolar sunspot group 37, *37*, 38, *42*, 43
Birmingham 84
"Black body" distribution curve 26, *26*, 28, *28*
"Black body" radiation 26, *26*, 28, *28*
Black dwarf 87
Black hole 83
Bohr, Niels *26*
Boron 82, *82*, 83
Bow shock 78
Brackett series 27, *27*
Brahe, Tycho *16*, 17, *17*
Bremsstrahlung radiation 67
Brightness temperature 28, 44
Bright points 45, 69, 71, 74
Bristlecone pines *81*
British Astronomical Association 90
Brookhaven National Laboratory 82, 83, *83*
Bunsen, Robert 19
Bursts 68
"Butterfly" diagram 40, *41*

## C

Calcium 29, *34*, 37, *50*, *65*, 66, 72, *86*
Carbon *15*, 82, 86
Carbon-14 80, *81*
Carrington, Richard 18
  and flare of 1859 66
  and "Spörer's law" 40
  rotation system 38
  rotation 1602 77
Cassini, G. D. 17
Celestial equator 10, *10*, 11
  and solar axis 39
Celestial sphere 10, *10*, 11
Cepheids 84
Chaldeans 16
China 16, 17, 80
Chlorine-37 83
Chromosphere 14–15, 32–35, *71*
  layers *34*, *52*, *53*, *65*
  spicules rising from *33*
Chromospheric network 32–33, *51*, *52*, *53*, *65*
  and plages 44
  and supergranules 31
  outlined by spicules *33*, 35
Climate 80–81, *81*
Comets 8, 76
Continuum *See* Spectrum
Convective zone 14, *14*
  and chromospheric heating 35
  and magnetic fields 43, *43*
  and solar oscillations 84
  and supergranules 31, *31*
Copernicus, Nicolaus 16, 17
Corona 15, *49*, *52*, *53*, *61*, *65*, 70–73, *70–73*, 74–75, 77
  mentioned by Plutarch 18
  observation of 20–21, *21*
  ultraviolet and X-ray radiation from 29

Coronal holes *53*, *60*, *62*, *63*, *64*, *65*, 74–75, *74*, *75*
  and chromosphere 32
  and solar wind 77
Coronal loop *44*, 68, 70, 71, *71*, 74
Coronal rain *47*, 48
Coronal transient 74–75, *74*, *75*
"Coronium" 72
Cosmic rays 79, 80
Cosmic year 6
Culgoora radioheliograph 21, 22, *22*, *48*, 74

## D

Daguerrotype process 19
Dark flocculi 46
Declination 10, *10*
De la Rue, Warren 19
Density
  mean, of the Sun 14–15, *15*, 82, 84
    compared with planets 7
    compared with stars 6
  of chromosphere 35
  of corona 15, 73
  of coronal hole 74
  of prominences 46
  of solar wind 76
*De Revolutionibus Orbium Coelestium* 16, 17
Deslandres, H. 20
Deuterium *82*
Diameter
  of the planets 8
  of the Sun 7, 8, *8*, *15*, 84, 85, *85*
"Diamond ring" effect *13*
Dicke, R. H. 84
Diffraction grating 19, 20
Differential rotation
  discovery of 18
  effect on magnetic field 42–43, *43*
  revealed by sunspots 39, *39*
Doppler shift 31, *31*
  and Evershed effect 36
  and rotation of photosphere 39
  and solar oscillations 84–85
  effect on spectral lines 27, 72, 73

## E

Earth (physical data) *9*
Earthquakes 84
Eccentricity 8, *9*
Eclipses 12–13, *12*, *13*
  and flash spectrum 34, *34*, *50*, *65*
  Baily's beads 85, *85*
  earliest record 16
  table of 92
Eclipse year 12
Ecliptic 10, *10*, 12, *12*, 39
Eddington, Sir Arthur 13
Edison, Thomas 21
Effective temperature 26
Egypt *5*, 16
Einstein, Albert 13
  *See also* Relativity
Electromagnetic spectrum *See* Spectrum

Electrons
  and aurorae 80, *80*
  and corona 73
  and flares 68
  and ionization 27
  and nonthermal radiation 28
  and proton-proton chain 82, *82*
  and solar wind 15, 76
  and white dwarf 87
  in Van Allen belts 79
  orbits *26*, 27
Elements
  and nuclear reactions 82–83
  and solar composition 6
  and spectral analysis 27
Ellipse 8, *9*, 12–13, 29
Emission line 26–29
  in flare spectrum *67*
  in spectrum of corona 72
Energy
  and luminosity 7
  generation of 6, 14, 35, 82–83, *82*
  level, atomic 26–27
Epicycle 16
Equation of time 11, *11*
Equinox 10, *10*, 11
Eros 17
Eshleman, V. R. 17
Euler, Leonhard 17
European Space Agency (ESA) 22
Evans, J. W. 46
Evershed effect 36
Evolution of the Sun, 7, 8, 86–87, *87*
Explorer spacecraft 22, 78
Extreme ultraviolet 26, 29, *63*, *65*
  and corona 70, *73*
  and flares 66, 67

## F

Fabricius *18*
Faculae *19*, 31, *31*
  "chromospheric" 44
  discovery of 18
  polar 45
Fan 70
"Fast drift bursts" 68
Fibril 33, 37, *44*
Filament 34, 36, 44, *44*, 46–48, *46–48*
  and flares 67, 68
  *See also* Prominence
Fitzeau, H. 19
Flamsteed, John 40
Flare 45, 66–69, *66*, *67*, *69*
  and coronal transients 74–75
  and "neutral line" 37
  and prominences 48
  and sunspots 37, 38, 43
  and X-rays 29
  first observation of 16
Flare star 69
Flash spectrum 34, *34*, *50*, *65*
Flocculi 46
Flux 42, 43, *43*, 44, 45, 69, 73
"Forbidden lines" 72, 73
Foucault, L. 19
Fraunhofer, Joseph von 19, *20*
Fraunhofer lines *20*, 34, 46, *50*, *65*, 72, 73
f-spot *See* Sunspots

The publishers gratefully acknowledge the assistance of Peter Gill in the preparation of this book. The table of solar eclipses on page 92 is reproduced, with permission, from the *Handbook of the British Astronomical Association* for 1973.

**Photographic Credits**
Cover: U.S. Naval Research Laboratory
p. 5(1) Kitt Peak National Observatory
p. 5(3) The Mansell Collection
p. 6(1) Lund Observatory, Sweden
p. 12(5A, 5B, 5C, 5D) Jay M. Pasachoff
p. 16(3A, 3B, 3C) Ann Ronan Picture Library
p. 17(4, 5) The Mansell Collection
p. 17(6A) Giraudon
p. 17(6B) Ann Ronan Picture Library
p. 18(1 & 2) Ann Ronan Picture Library
p. 18(3) The Science Museum (from frontispiece of Scheiner's *Rosa Ursina Sive Sol*, 1630)
p. 19(4) Ann Ronan Picture Library
p. 19(5 & 6) Royal Astronomical Society
p. 20(2A, 2B) Mount Wilson and Las Campanas Observatories, Carnegie Institution of Washington
p. 22(2) CSIRO Solar Observatory/Photo by Tamair, Tamworth N.S.W.
p. 23(4) Project Stratoscope, supported by NASA, NSF & ONR (Courtesy of Princeton University)
p. 23(5) U.S. Naval Research Laboratory
p. 27(3) Kitt Peak National Observatory
p. 29(3) Max-Planck-Institut Für Radio Astronomy
p. 29(5A, 5B, 5C, 5D) Solar Physics Group, American Science & Engineering, Inc.
p. 30(1) Royal Greenwich Observatory
p. 30(2) CSIRO Solar Observatory, Culgoora, N.S.W.
p. 31(4) Courtesy of R. Leighton, Mount Wilson and Palomar Observatories, Carnegie Institution of Washington
p. 31(5) Observatoire du Pic-du-Midi
p. 32(1) Association of Universities for Research in Astronomy, Inc., Sacramento Peak Observatory
p. 32(2) Sacramento Peak Observatory, Air Force Cambridge Research Laboratories
p. 34(1) Lick Observatory
p. 34(2A, 2B) Mount Wilson and Palomar Observatories, Carnegie Institution of Washington
p. 35(2C, 2D) Mount Wilson and Palomar Observatories, Carnegie Institution of Washington
p. 35(3) CSIRO Solar Observatory, Culgoora, N.S.W.
p. 36(1) Mount Wilson and Las Campanas Observatories, Carnegie Institution of Washington
p. 37(2A, 2B, 2C) The Aerospace Corporation, San Fernando Observatory
p. 37(3) CSIRO Solar Observatory, Culgoora, N.S.W.
p. 38(1) Kitt Peak National Observatory
p. 39(6A, 6B, 6C) Plate 3.25 from the book *Sunspots* by R.J. Bray and R.E. Loughmead (Dover Publications, Inc., NY, 1979)
p. 40(1A, 1B) Astronomical Institute of Czechoslovakian Academy of Sciences, Observatory Ondřejov
p. 41(3) Royal Greenwich Observatory
p. 42(1A, 1B) Kitt Peak National Observatory
p. 44(1A, 1B, 1C) Solar Physics Group, American Science & Engineering, Inc.
p. 44/45(2A, 2B) Big Bear Solar Observatory, Caltech
p. 46(1A, 1B, 1C) Observatoire de Paris—Meudon
p. 46(2) Astronomical Institute of Czechoslovakian Academy of Sciences, Observatory Ondřejov
p. 47(4) Association of Universities for Research in Astronomy, Inc., Sacramento Peak Observatories
p. 47(5) Big Bear Solar Observatory, Caltech
p. 47(6) Sacramento Peak Observatory

p. 48(1A) Mees Solar Observatory (University of Hawaii)
p. 48(1B, 1C, 1D, 1E) CSIRO Solar Observatory, Culgoora, N.S.W.
p. 48(2A, 2B, 2C, 2D) Observatoire de Paris—Meudon
p. 49 William P. Sterne, Jr.
p. 50(top) Association of Universities for Research in Astronomy, Inc., Sacramento Peak Observatories
p. 50(bottom) Dennis de Cicco/*Sky and Telescope*
p. 51/52/53 Dr. R. Levine, Harvard College Observatory
p. 54(top) Dr. R. Levine, Harvard College Observatory
p. 54(bottom) NASA
p. 55 NASA
p. 56/57 NASA/Harvard College Observatory/E.J. Schmahl, UMd.
p. 58 Dr. R. Levine, Harvard College Observatory
p. 59(top) U.S. Naval Research Laboratory
p. 59(bottom) Solar Physics Group, American Science and Engineering, Inc.
p. 60 Marshall Space Flight Center
p. 61 High Altitude Observatory, N.C.A.R.
p. 62 Dr. R. Levine, Harvard College Observatory
p. 63 NASA
p. 64 Solar Physics Group, American Science and Engineering, Inc.
p. 66(1A) Big Bear Solar Observatory, Caltech
p. 67 all except for top center left from Aerospace Corporation, San Fernando Observatory
p. 67 (top center left) U.S. Naval Research Laboratory
p. 68 Dr. Sara F. Martin/Lochkerd Solar Laboratory
p. 70(1) High Altitude Observatory, N.C.A.R.
p. 71(2A) A. Wallenquist, Uppsala Observatory
p. 71(3A) Sacramento Peak Observatory of the Air Force
p. 71(4) High Altitude Observatory, N.C.A.R.
p. 71(5) Sacramento Peak Observatory, Air Force Cambridge Research Laboratories
p. 72 High Altitude Observatory, N.C.A.R.
p. 73(2) William E. Behring, Laboratory for Astronomy & Solar Physics, Goddard Space Flight Center
p. 73(4) Solar Physics Group, American Science and Engineering, Inc.
p. 74(2A, 2B) Solar Physics Group, American Science and Engineering, Inc.
p. 75(3) High Altitude Observatory, N.C.A.R.
p. 75(4A, 4B, 4C, 4D, 4E, 4F) Solar Physics Group, American Science and Engineering, Inc.
p. 75(5) U.S. Naval Research Laboratory
p. 80(1) NASA
p. 85(bottom right) J.H. Mathers
p. 86 Courtesy of the Royal Astronomical Society

**Artwork credits**
p. 6(2) From "The Milky Way Galaxy" by Bart J. Bok. © (1981) by Scientific American, Inc. All rights reserved.
p. 28(1) based upon data from *The Solar Output and its Variations*, edited by O.R. White (Colorado Associated University Press, Boulder, 1977)
p. 38(2) Pub. Eidg. Sternwarte Zürich, vol. IX, No. 1, 1947, by M. Walmeier (ed.)
p. 39(3) From "The Rotation of the Sun" by R. Howard. © (1975) by Scientific American, Inc. All rights reserved
p. 71(2B) A. Wallenquist, Uppsala Observatory
p. 71(3B, 3C, 3D) From "Ultraviolet Astronomy" by Leo Goldberg. © (1969) by Scientific American, Inc. All rights reserved.
p. 76(1, 3A, 3B, 4A, 4B, 5) From "Streams, Sectors and Solar Magnetism" by Arthur J. Hundhausen (in *The New Solar Physics*, ed. J.A. Eddy)
p. 81(2, 4) From "The Case for the Missing Sunspots" by J.A. Eddy. © (1977) by Scientific American, Inc. All rights reserved.

**Illustrators**
Marilyn Bruce, Chris Forsey, Mick Gillah, Colin Salmon, Mick Saunders